Understanding Automation Systems

Written by: Neil M. Schmitt, Ph.D.
Dean of Engineering
Prof. of Electrical Engineering
University of Arkansas
Staff Consultant, Texas Instruments Information Publishing Center

Robert F. Farwell, BSEE, MBA
Technical Sales Representative
Norcom Inc., Dallas, Texas
Staff Consultant, Texas Instruments Information Publishing Center

With contributions by: Gerald Luecke, MSEE
Mgr. Technical Products Development
Texas Instruments Information Publishing Center

Charles W. Battle, Editor
Leslie E. Mansir, Editor
Texas Instruments Information Publishing Center

TEXAS INSTRUMENTS

P.O. BOX 225012, MS-54 • DALLAS, TEXAS 75265

This book was developed by:

The Staff of the Texas Instruments Information Publishing Center
P.O. Box 225012, MS-54
Dallas, Texas 75265

The authors wish to acknowledge:

Marketing and Engineering Staff of T.I. Johnson City, TN.
  Mike Bradley for his consultation on initial concept and outline, guidance
  and support; Bill White for his hospitality and responses to numerous
  requests for information; Jeff Hadley for his technical review; Tom Welch,
  Rob Kleinsteuber, Shannon Rhea and Teresa McClain for their assistance.

Engineering Staff of T.I. Dallas, TX.
  Bill Rylander and Clyde Golightly for their technical contributions and
  guidance, Gib Hagler and Chris Winemiller for their assistance.

Bob Farwell expresses special appreciation for the patience and continued
support that he received from Bob Medlin, Ken Norvell and his wife, Winnie.

Word Processing:	*Design and artwork by:*
Betty Brown	Plunk Design

ISBN 0-672-27014-5
Library of Congress Catalog Number: 84-51472

Second Edition
Second Printing

About the cover:

Represented on the cover are the components of modern day automation
systems: a robot end effector or "hand" positioning an integrated circuit for
insertion in a printed wiring board, the robot controller and detail drawings of
the axes of movement of the robot arm.

Table of Contents

Preface

This book is about automation — specifically the electronic control of automated processes. This book is also about productivity, competition and profit as these are affected by decisions to employ or ignore automating a business or industry. The second industrial revolution is upon us and those who will survive the competition must enter into the arena of change.

The foundation of the second industrial revolution is the availability of complex computing power at low cost, high reliability, small size, low power and high performance. This foundation has been made possible by the rapid and profound advances in microelectronics, semiconductor integrated circuits and computer technology.

This book is not a design manual. It is a book written to introduce the reader to the subject of electronic control of automation systems and to help the reader understand the terms, principles, techniques and effort used to automate processes. It begins with an understanding of the basic electronic and control concepts used in automation systems. Continuous, discrete and batch processes each are discussed by way of examples. Next, the actual mechanical hardware used in controlling an automated system is shown. After a description of the electronic components found in control systems, particular digital circuits and the combination of these circuits to make a functional computer are discussed. A discussion of computer software and programming languages concludes the introduction of basic concepts.

Much of the remaining portion of the book is devoted to applications of the three types of processes commonly encountered — continuous, semi-continuous and discrete parts. Several system examples of each are included which build on the concepts previously introduced. Since robots are playing an increasingly important role in automation systems, a chapter is dedicated to the explanation of their function, to the types available and to typical applications. The book concludes with an example of electronic control of an automation system that ties all of the topics in the book together and with a discussion of the future trends in automation.

Like other books in this series, this book builds understanding through the subject in a step-by-step manner. Knowledge and confidence are gained as each subject is completed. Readers experienced in the fundamentals of electronics and control could begin at Chapter 5, but the material contained in Chapters 1 through 4 forms an excellent basic review. A quiz is included at the end of each chapter for self-evaluation of what has been learned. Answers to the questions are found in the back of the book.

Increased productivity is vital for companies to maintain their competitive position in the world market. Electronic control of automation systems is the key. We sincerely hope that this material will prove useful to help you make the critical decisions of the future.

N.S.
R.F.

History and Importance of Industrial Control

INTRODUCTION

The National Academy of Sciences recently stated: "The modern era of electronics has ushered in a second industrial revolution ... its impact on society could be even greater than that of the original industrial revolution." Henry Ford championed the industrial revolution just after the turn of the century with his Model T Ford and assembly line manufacturing. Whether a person is part of the automobile industry today, or in manufacturing of any type, or in commerce, banking, insurance, transportation, education, or in most any field, the impact of electronics is seen and felt daily. In addition, it seems the pace of the impact is moving more rapidly each day. Probably the main reason for this occurrence is that electronics can help improve productivity (output per man-hour) and keep our U.S. goods competitive worldwide. Not only is productivity improved, but also the product that is being manufactured or the task being performed is improved by better performance, reduced power consumption, better reliability and lower cost. That's what makes electronic control of automation systems so exciting.

ABOUT THIS CHAPTER

The goal in this chapter is to provide you with the motivation and desire to understand the principles involved in the electronic control of automation systems so that you may be a participant in this exciting era rather than a spectator. Automation is defined and its effect on industry and society is discussed. Also introduced are the basic types of electronic control devices and systems which will be discussed in more detail in the following chapters.

AUTOMATION

What Is It?

Automation is the control of automated processes to gain accuracy or precision, but above all, to gain productivity.

This question can generate a variety of responses from people because they tend to define automation as a function of their environment and experience. To the housewife it may be the automatic dishwasher; to the factory worker it may mean a robot; to the company president it could be the difference between profit and loss. In the broadest sense, automation is the control of automated processes. In research and laboratories, automation may be used for accuracy or precision, but in most industrial applications, automation is used to increase productivity.

All of these concepts include the idea of using electrical and/or mechanical power to drive some kind of machine. Added to the machine is some amount of "intelligence" that controls the machine to do its task so that productivity is increased, costs are reduced or both. The machine can operate longer hours in a harsher environment and perform some tasks more rapidly with less error than its human counterpart. Further, the machine never complains, never strikes, and doesn't ask for a raise, vacation or increased fringe benefits. On the other hand, the machine has very limited decision making capability compared to humans and must have predefined instructions (a program) to control its operation in every situation.

Part of this second industrial revolution, then, has been to place the worker in control of the machines instead of actually performing the task. Thus, the housewife must learn to load the dishwasher properly and be aware of its limitations, rather than learn to wash the dishes themselves. Similarly, the milling-machine operator must be willing and capable of being trained to set up and operate a numerically-controlled machine that performs the actual milling of the part.

Much of the "second industrial revolution" is to relieve the worker from performing a task directly by providing controllers of machines to do the task.

With the advent of the integrated circuit (IC) in the early 1960's and the microprocessor in the early 1970's, the amount of intelligence that could be built into a machine at a reasonable cost took a giant leap forward. The number of complex tasks that could be automated increased several fold. It is now economically possible to dedicate small computers (microcomputers) to the performance of a single task.

How Does it Affect Jobs?

Many people have an unfounded fear that "automation" means "loss of jobs", when just the opposite is often true. In fact, *lack of automation* may put people out of work. If companies cannot compete economically because of lower productivity due to lack of automation, they will be forced to lay off personnel or even close down.

Automation does not mean job loss, but rather job gain and stability because of the increase in productivity, efficiency and savings.

This has been dramatically illustrated in the automobile industry and the steel industry. In 1981-82, the U.S. automobile industry laid off over 40,000 management and 250,000 union workers because of loss of sales. The cost of their product had increased so that the average person could no longer pay the price. Other competitors were giving more value for the price.

The steel industry was much the same story. However, while most U.S. steel mills were laying off workers due to lack of sales, a Texas mill that was fully automated was expanding and had a backlog of orders because it had an economic advantage.

Many applications of automation do not involve replacement of people because the function did not even exist before. One company was able to save thousands of dollars per year by automatically monitoring and controlling the amount of oxygen in the flue gas of its steam boilers for most efficient fuel consumption. An Arkansas firm realized a $300,000 annual savings by implementing an automated system to recover acetone from an airsteam vented to the atmosphere. In both of these cases, air pollution was reduced along with the cost savings.

WHERE DOES THE U.S. STAND?

The United States has slipped behind foreign competitors in productivity and market share due to lack of aggressive automation of its industries.

Consumers in the United States generally purchase goods that offer the best quality for the least money whether or not they were manufactured in the U.S. The success of foreign manufacturers in the automotive, photography, entertainment, electronics and clothing markets indicates the intensity of the competition. Over the last 20 years, the sales of U.S. manufacturers have fallen farther and farther behind the sales of foreign companies in most cases. A main reason has been less value for the price, and the price has increased because of lower productivity. In the last two decades, productivity in the United States has risen only 1.6 percent per year while it has risen in Japan at a rate of 7 percent per year; in Italy at 4.5 percent per year; and in France at 4.1 percent per year. Lower productivity means higher manufacturing costs and, because of the slow increase in productivity, manufacturing costs have risen faster in the U.S. than in these other countries; thus, U.S. prices are higher. But why has U.S. productivity slipped?

Productivity is affected by many complex and inter-related factors; automation is a major contributing factor for increasing it. Other countries often have been more aggressive in automating than the U.S.; for example, there are approximately 25,000 industrial robots currently in use worldwide and over half of these are in Japan.

WHY IS THE U.S. BEHIND IN AUTOMATION?

Why is the U.S. behind in the application of automation to industrial processes when most of the fundamental techniques and equipment for automation were developed in the United States? For example, the transistor was invented at Bell Laboratories. The integrated circuit (IC) was invented at Texas Instruments. The first large- and medium-scale digital computers, minicomputers, microcomputers, and microprocessors were all developed in the United States. Early work on remote manipulators to handle radioactive substances led to the first industrial robots in the U.S. before 1970. The U. S. Air Force funded the first programmable numerical control systems for machine tools required to machine the skins, spars, and other members necessary to build the structures of the military's high-performance aircraft.

Even though the U.S. pioneered the technology that is used in automation, foreign competitors have used it quicker and more efficiently than many U.S. industries.

Process control equipment was almost a U.S. monoploy—even that sold overseas came mainly from subsidiaries of U.S. companies or through companies licensed by U.S. companies. Yet the U.S. companies now find themselves in a peculiar position because their foreign competitors have put the equipment to work sooner and more efficiently than they have.

Let's again use the automotive industry, the largest single-product industry in the world, as an example of the impact of automation. In 1960, the United States produced 48% of the world's motor vehicles (about 8 million) while Japan produced 3%. Further, the U.S. exported 4% of production while only 6% of sales in the U.S. were imports.

In 1980, the U.S. again produced about 8 million vehicles, but captured only 20% of the world market while Japan accounted for 28% of the world production. Also in 1980, 25% of sales in the U.S. were for imports. Loss of sales to U.S. companies exceeded $4 billion in 1980 alone. Other industries also are experiencing significant loss of sales to foreign competition.

HOW CAN THE U.S. CATCH UP?

Many people will argue that automation is not the cause of this loss of sales. It isn't the *only* cause, but what are U.S. manufacturers doing to meet this competition? *They are automating!* U.S. automakers spent more money in 1981 automating their plants than has been spent for the entire space shuttle program. Evidently they feel automation is the key to revival—or survival. The projected total investment by the auto industry will exceed the cost of the Apollo space program and will be completed in about one-half the time.

Other industries also have begun to move heavily into automation. This has created a big demand for industrial control equipment. The size of the market for equipment and services in the automation area has been estimated to be between $6 billion and a staggering $50 billion annually with a growth rate of 10 to 15 percent per year. Conservative estimates seem to indicate that $10 billion dollars is more reasonable, with one-half for equipment and the other half for services. Let's take a brief look at what kind of industrial control equipment makes up this market and what it does.

CONTROL—THE HEART OF AUTOMATION

Automated system control depends on the ability of the machines to compare, monitor and adjust to achieve a desired product with little or no human help.

Automation of an industrial process is dependent on the ability to control the process with little or no help from humans. What has to be controlled depends on the process, but generally control involves starting, stopping and regulating the movement, position or flow of each of the components of the process. The ability to control, in turn, usually depends on the ability to monitor or measure variables in the process that need to be controlled to ensure that the final product is as desired. This depends, in turn, on the ability to compare the actual product to the desired product and make adjustments in the process if the error exceeds some predetermined threshold. A system that has all these abilities is called a control system. Chapter 2 will develop these concepts in more detail.

ELECTRONICS—THE HEART OF CONTROL

Sensors that measure the status of important variables in a control system are the inputs to the system, but the heart of the system is the electronic controller. Most of the automation systems described in this book are possible only because of the recent dramatic advances in electronics. Control system designs that were not practical because of cost only five years ago are already becoming obsolete today because of the rapidly advancing technology.

Electronic control devices are the heart of any automation system.

As an example, consider the microcomputer. The single-chip microcomputer is a computer on a piece of silicon about 0.25 by 0.25 inch in size that fits on the end of a finger (*Figure 1-1*). The first electronic digital computer weighed over 60,000 pounds and filled several rooms. It required several thousand watts of power to operate and its maximum speed was less than 5,000 calculations per second. It cost millions of dollars. In contrast, the tiny microcomputer chip that can sit on a fingertip is approaching 1,000,000 calculations per second, operates on less power than a flashlight, and costs under $100.00.

**Figure 1-1.
Single-Chip
Microcomputer**

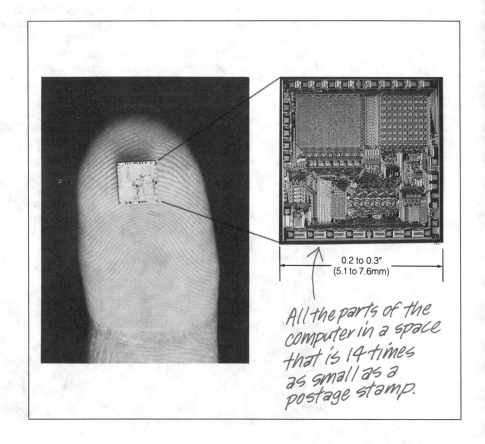

0.2 to 0.3"
(5.1 to 7.6mm)

All the parts of the computer in a space that is 14 times as small as a postage stamp.

The key to successful automation and competitive manufacturing today is to use high-technology electronics to provide flexible manufacturing through programmable electronic systems. Boeing Aircraft Corporation builds its commercial aircraft on an assembly line, but provides for unique interior designs by simply changing a computer program. Chrysler Corporation uses robot welders, as shown in *Figure 1-2*, to weld automobile bodies. Thirty robots nearly doubled the output (from 60 to 100 cars per hour) with significantly fewer reliability problems.

The inexpensive computer power now available has enabled these companies to automate and at the same time tailor the systems to their applications. To tailor the computer to the application, the set of instructions (called a program) that control the computer is organized for the application. If, for instance, the computer is controlling a robot that drills holes in a piece of sheet metal, the drilling pattern, depth of holes, etc. can be changed completely simply by changing the program of the computer. Usually the computer program is stored on an electronic device called a memory chip. Changing the program on this chip may require only fifteen minutes time. Of course, it does take some time to rewrite the program that is put on a memory chip; however, the time and cost involved is several times less than retooling a machine.

Chapters 3 and 4 of this book will discuss the sensors, actuators and electronic components commonly used in control systems.

The revolution in high technology electronics enables companies to design flexible and complex automated systems quickly and inexpensively.

Figure 1-2.
Robot Welders
(Courtesy of Chrysler Corporation)

AUTOMATION SYSTEMS

The application of electronic automation to industrial processes has resulted in various types of automation systems. These can be generally classified as follows:

1. Numerically Controlled Machine Tools
2. Programmable Controllers
3. Automatic Storage and Retrieval Systems
4. Robotics
5. Flexible Manufacturing Systems

Each type is briefly described here and will be discussed in more detail in succeeding chapters.

Numerically Controlled (NC) Machine Tools

The first type of automation system is a numerical controlled machine tool, in which 3 coordinates are controlled to position a tool that drills, grinds, shapes, or mills. Productivity gains of 3:1 are possible.

A machine tool is a power driven tool or set of tools that remove material from a workpiece by milling, shaping, drilling, grinding; or inserting parts on a workpiece. A machine tool can be controlled in one of two ways:

1. Continuous control of the tool path where work is continuous or nearly continuous on the workpiece.
2. Point-to-point control of the tool path where work is performed only at discrete points on the workpiece.

In either case, three coordinates must be specified to position the tool at the correct location. Computer programs exist for calculating the coordinates and for producing a paper or magnetic tape that contains the numerical data actually used to control the machine. Productivity gain through using NC machines is up to 3 to 1. Also, less operator skill is required and one operator may be able to operate more than one machine.

If instead of using a tape to control the machine, a dedicated computer is used, then the system is technically called a computer numerically controlled (CNC) machine. The CNC machining center shown in *Figure 1-3* is capable of selecting from over 20 tools and performing many different operations such as milling, tapping, etc.

If the computer is used to control more than one machine, the system is called a direct numerically controlled (DNC) machine. The advantage of this approach is the ability to integrate production of several machines into the overall control of an assembly line. The disadvantage is the dependency of many machines on the health of one computer. Chapter 7 will show examples of additional NC machines.

Figure 1-3.
CNC Machine Tool
*(Photo courtesy of Kearney &
Trecker Corporation)*

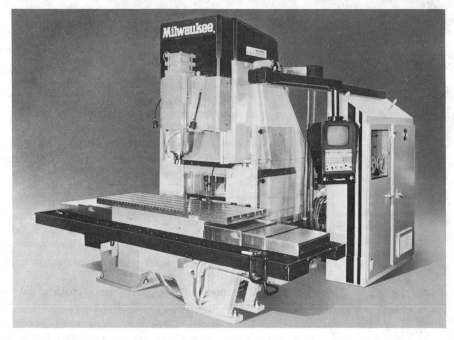

By incorporating in various combinations robotics, computers and NC machines, flexible manufacturing systems are feasible and will eventually lead to totally automated manufacturing systems.

Programmable Controllers

Programmable controllers are just what the name implies. They are solid-state devices that not only can control a process or a machine, but also have the capability of being programmed or reprogrammed rapidly as the need may arise. Often, a keyboard or numerical keypad may be connected to the controller for programming. *Figure 1-4* shows a typical programmable controller.

Figure 1-4.
Programmable Controller

The second type, programmable controllers, are less complex but easier to design and install and are less expensive.

These devices vary in the complexity of operation they can control, but they can be interfaced with a minicomputer or microcomputer and operated as a DNC machine in order to increase their flexibility. On the other hand, they are relatively inexpensive and easy to design and install. Chapters 6 and 8 discuss applications of programmable controllers.

Automatic Storage and Retrieval Systems

The third type, storing and retrieving items automatically, has a computer controlling fork lifts, cranes and similar conveyances.

Warehousing activities center around storing an inventory of parts or materials for later retrieval and shipping or use. In automated systems, a remotely located computer controls fork lifts, cranes or similar conveyances to receive, store, and retrieve inventory items. Inventory control is accurate and items can be shipped or used according to the date received. The McDonald Restaurant chain has an automated warehouse for storing frozen french fries. The Wal-Mart Discount Center Headquarters, where hundreds of inventory actions occur daily, uses an automated warehouse for central collection and distribution. These are but a few of the exciting new applications in this area.

Robotics

A robot is a computer controlled device capable of movement in one or more directions while performing a sequence of operations. A CNC machine can be considered a robot, but usually the use of the term robot is restricted to those devices having movements like those of humans, particularly the arm and hand movements.

The fourth type, robotics, usually is an automated device that has some motion characteristics similar to humans.

The tasks they perform may be machining tasks such as drilling, but usually are tasks such as welding, picking and placing, assembling, inspecting, and painting. We normally think of industrial robots as massive devices like those shown in *Figure 1-2*, but actually they may be quite small. Many applications can use robots that mount on a bench or table like the one shown in *Figure 1-5*.

If a task is relatively simple, repetitive, or would be hazardous to a human, then a robot may be an appropriate choice. Robots are increasing in intelligence with the addition of sight and hearing and this will allow more complex tasks to be performed by robots. Robots are so important to the future of manufacturing systems that all of Chapter 9 is devoted to exploring their potential.

Flexible Manufacturing Systems

The fifth type of automation system is a combination of all the rest of the types to provide a flexible manufacturing system for all types of industry.

The incorporation of NC machines, robotics and computers into an automated assembly line results in what is called a flexible manufacturing system. It is called flexible because major changes can be made with relatively little investment of time and money. In its ultimate form, raw materials will enter at one end and the finished product will exit from the warehouse at the other end for shipping without human intervention. Today this exists only in concept, although major portions of such a system are already in place. However, with advances in automation systems such visions will soon be reality.

Figure 1-5.
Small Robot Arm
*(Courtesy of Copperweld
Robotics)*

WHAT HAVE WE LEARNED?

1. An industrial revolution involving automation of manufacturing processes is taking place.
2. Automation is the use of electrical and/or mechanical power controlled by an intelligent (usually electronic) control system in order to increase productivity and reduce costs.
3. Lack of automation may increase unemployment.
4. Other countries are increasing productivity faster than the United States and are capturing larger shares of the product market.
5. Automation is one way to increase productivity in the United States.
6. The ability to control the steps of a process is the key to automation.
7. Advances in electronics have made control of complex systems possible at a cost-effective price.
8. The several kinds of automation systems that can be applied to industrial processes are: numerically controlled machines, programmable controllers, automatic storage and retrieval systems, robotics and flexible manufacturing systems.

Quiz for Chapter 1

1. The new industrial revolution has been caused by:
   a. social unrest.
   b. advances in electronics.
   c. demand for higher wages.
   d. the Vietnam War.

2. Automation means:
   a. workers controlling machines.
   b. assisting or replacing humans with machines.
   c. increased productivity.
   d. all of the above.

3. Automation normally causes:
   a. a net loss of jobs.
   b. a net increase in jobs.
   c. no change in jobs.
   d. none of the above.

4. Compared to humans, machines:
   a. can work in harsher environments.
   b. can make more complex decisions based on unexpected circumstances.
   c. make fewer errors.
   d. a and c.

5. Automation usually:
   a. displaces some workers.
   b. increases productivity or profit.
   c. gives a competitive advantage.
   d. all of the above.

6. Productivity is defined as:
   a. number of items manufactured per day.
   b. output per man-hour of labor.
   c. cost per unit.
   d. cost per day.

7. In general, American buyers will purchase goods:
   a. that are made in America, even at higher costs.
   b. that offer the best quality for the least money.
   c. without regard to country of manufacturer.
   d. b and c.

8. Over the last ten years, productivity has risen *most* rapidly in:
   a. Italy.
   b. Japan.
   c United States.
   d. France.

9. Over the last ten years, productivity has risen *least* rapidly in:
   a. Italy.
   b. Japan.
   c. United States.
   d. France.

10. Comparing 1980 to 1960, automobile production in the United States has:
    a. risen.
    b. fallen.
    c. stayed about the same.

11. From 1960 to 1980 the U.S. share of the world auto market has:
    a. increased about 10%.
    b. decreased about 28%.
    c. decreased about 17%.
    d. increased about 39%.

12. In order to combat foreign competition, U.S. companies are:
    a. automating.
    b. seeking import tariffs.
    c. ignoring the situation.
    d. decreasing salaries.

13. Use of modern control technology in automation systems:
    a. increases yield.
    b. reduces costs.
    c. improves reliability.
    d. all of the above.

14. The heart of automation technology is:
    a. robots.
    b. computers.
    c. sensors.
    d. control.

15. The electronic component that has been the key to recent developments in automation is the:
    a. improved vacuum tube.
    b. integrated circuit.
    c. transistor.
    d. diode.

16. A microprocessor is:
   a. an integrated circuit.
   b. controlled by a program.
   c. relatively inexpensive.
   d. all of the above.

17. Programmable devices are
   advantageous because:
   a. changes can be made quickly.
   b. microprocessors are then
   unnecessary.
   c. only highly skilled professionals
   can make changes.
   d. none of the above.

18. The largest user of robots is:
   a. United States.
   b. Soviet Union.
   c. France.
   d. Japan.

19. Which of the following may be
   classified as an automation system?
   a. numerically controlled machine
   tools.
   b. automated warehouses.
   c. robotics.
   d. all of the above.

20. A flexible manufacturing system
   may be:
   a. an automated assembly line.
   b. very difficult to change when new
   products are introduced.
   c. expensive to alter.
   d. all the above.

Industrial Control Fundamentals

ABOUT THIS CHAPTER

After reading the first chapter, you should realize that the application of electronic control concepts to manufacturing and other processes is both desirable and necessary. In this chapter, we will discuss the kinds of processes to be controlled, the types of systems, and the methods available to perform the control functions. The components and signals commonly present in control systems will be covered. Open-loop and closed-loop control systems will be defined and the fundamental characteristics of control systems will be discussed.

KINDS OF PROCESSES

Industrial and manufacturing processes can be grouped into three general areas in terms of the kind of operation that takes place:
1. Continuous
2. Batch
3. Discrete Items

Each of these kinds of processes have unique characteristics and present different challenges to designers of electronic control systems.

Continuous Process

In a continuous process, material moves continuously from raw material to finished product.

A continuous process is one where raw materials enter one end of the system and the finished product comes out the other end of the system as indicated in *Figure 2-1* while the process itself runs continuously. For such an application, continuously means a relatively long period of time. The time period may be measured in minutes, days, or even months depending on the process.

**Figure 2-1.
A Model of a Continuous System**

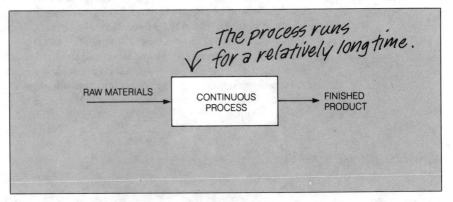

The process runs for a relatively long time.

RAW MATERIALS → CONTINUOUS PROCESS → FINISHED PRODUCT

Consider the steel rolling process shown in *Figure 2-2*. Long pieces of sheet steel are moved along a conveyor system through sets of pressurized rollers that squeeze the steel to reduce the thickness of the sheet. This is a continuous process that is measured in minutes depending on the length of the sheet. The steel sheet may move through the rollers at over 1,500 feet/minute in some cases. The loopers between the pressure rollers must keep constant tension on the sheet steel. In the steel rolling process, the conveyor speed, roller speed and pressure, and looper tension must be controlled. In particular, the load on the first set of rollers is subjected to sudden changes when new sheets of steel enter the rollers.

**Figure 2-2.
Sheet Steel Rolling
Process**

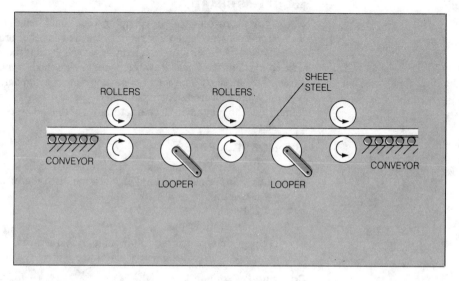

Batch Process

In batch processing, a given quantity of material is processed through its manufacturing steps as a unit, each step being completed before the unit passes on to the next step.

In a batch process, a set amount of each of the inputs to the process are received in a batch, then some operation is performed on the batch to produce a finished product or an intermediate product that needs further processing.

Figure 2-3 illustrates a batch processing system. Coffee beans are unloaded from barges and stored in silos. As beans are needed for processing, they are moved by suction through 6-inch diameter ducts (called airveying) to a weigher which creates a 1,200 pound batch of beans. This batch then is moved through the steps of the process as a batch, not as a continuous flow. After roasting, the beans are water quenched and airveyed to other bins for blending, grinding and packing. At least six separate batch processes can be seen here: weighing, roasting, quenching, blending, grinding, and packing. Each of these steps do something unique to the coffee bean, then pass it on to the next step.

**Figure 2-3.
Batch Processes in
Coffee Making**

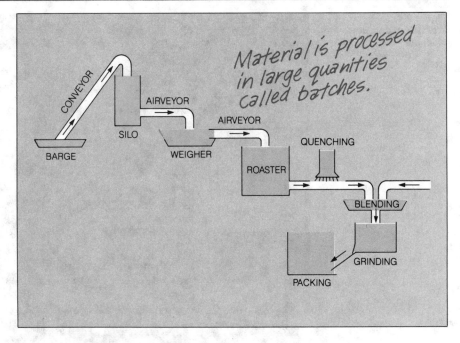

At each step, there are several things to control. For instance, consider the roasting process. The roaster is a perforated barrel that is loaded with a batch of beans. Hot air from a furnace is pulled through the beans by a fan while air flow, temperature, and moisture content are measured. Air flow and temperature are controlled and the roasting process continues until the measured moisture content indicates that the beans are sufficiently dried. The controlling process at this step can be considered a continuous process, even though the total process is a batch process.

Discrete Parts Manufacturing

In discrete processing, each item to be manufactured is processed at each step as a separate, individual item. It is the most common processing system.

Of all the processing systems, the discrete parts manufacturing is the most common. In this manufacturing process, a series of operations, many of them similar, produces a useful output product. This kind of process is different from batch processing because, as in the coffee example, outputs of batch processing normally move on to additional batch processing before an end product is obtained. The workpiece normally is a discrete part that must be handled on an individual basis.

As an example of a discrete parts process, consider the double end trim and bore machine shown in *Figure 2-4*. A wood piece is placed in the machine and the operator presses a foot-operated switch to begin the process. The wood piece is clamped into place and two circular saws are lowered to trim both ends of the piece. The saws retract and two drills bore both ends to preset, but different, depths. The drills retract and the wood piece is released and removed. *Figure 2-5* shows a close-up of the clamp, saw, and drill. From this station the piece moves on to another operation such as sanding and sealing.

**Figure 2-4.
Wood Trim and Bore
Machine**
*(Courtesy of Colonial Wood
Products)*

**Figure 2-5.
Closeup of Clamp, Saw,
and Drill**
*(Courtesy of Colonial Wood
Products)*

COMPONENTS OF A CONTROL SYSTEM

In Chapter 1, we mentioned that a control system is the heart of an automation system. The major components of a control system are shown in *Figure 2-6*. They are divided into the general functions that must occur in the system. Further detail of what the general components contain will be provided as the book continues.

Figure 2-6.
General Control System

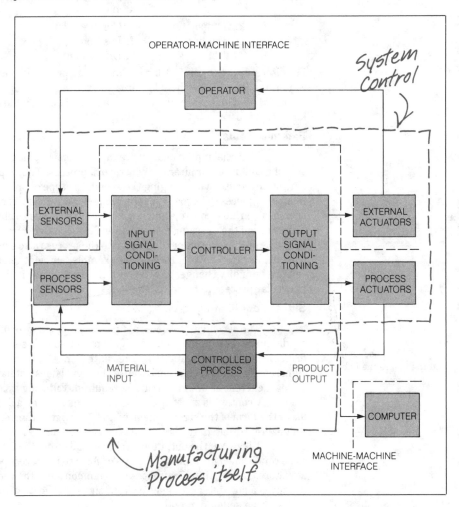

Sensors

The first of these functions is to recognize inputs from the process and from the external environment with sensors. Sensors are a type of transducer which convert physical information such as pressure, temperature, flow rate, position, etc. into electrical signals. The electrical signals are related to the physical variable in a known way so that the electrical signal can be used to monitor and control a process.

Sensors are usually categorized by what they measure. For continuous processes, one or more of these quantities often are needed:

1. Flow
2. Temperature
3. Pressure
4. Velocity
5. Acceleration

For batch processing, it may be one or more of these quantities:

1. Level
2. Composition
3. Weight
4. Volume
5. Tension
6. Compression
7. Position
8. Dimension

For discrete parts manufacturing, the sensors primarily measure ON/OFF conditions. The reader should realize that the above listing is general and that a sensor listed under one process very easily could be found in any of the other kinds of processes.

External Inputs

External inputs usually are inputs from a human to set up the starting conditions or alter the control of a process. They may include emergency shutdown, changing the speed, the type of process to be run, the number of pieces to be made, or the recipe for a batch mixer. Various types of switches, variable controls and keypads often are used to allow human inputs. This is called the operator-machine interface. Temperature, pressure and humidity of the environment external to the process also can be monitored and used to alter the control process. For example, ambient temperature and relative humidity changes may require the drying time and temperature of a particular process to be adjusted.

Signal Conditioning

The second of the general functions is some form of signal conditioning. The input electrical signal produced by a sensor often is not in a form that can be used directly by the controller. Amplification is a common signal conditioning technique because the signal does not have the right voltage, current or power amplitude required. Filtering a signal to change its frequency content is another common technique. When a digital controller is used, the form of the signal may need to be changed from a continuous signal to a signal represented by digital levels or vice versa.

Likewise, the output electrical signals from the controller often need to be conditioned in amplitude, shape or form to properly operate the actuators. Again, amplification is very common since the controller output usually is a low voltage and low current signal, while actuators usually require high voltage and/or current.

Actuators

Actuators convert system output electrical signals into physical action.

The third general function is that of producing output actions for the process and externally. Actuators are the devices that convert an electrical control signal to a physical action. For continuous processes, actuators usually are flow control valves, pumps, and positioning drives. For batch processes, they include variable speed drives, clutches, and brakes. For discrete parts manufacturing, they may include solenoids, stepping motors, stepping switches, and power relays. Again, these are general listings because any actuator may be used in any type of process.

External Outputs

Usually it is desirable that the controller indicate the state of the process or the value of certain process variables to the human operator. The external outputs to do this may be a meter, CRT (cathode ray tube) screen, printer, bell, light or other device with which the operator can interact. This is a machine-operator interface. Outputs also could go directly from the controller to a computer for storage of data and analysis of results. This is a machine-machine interface.

Controller

The controller makes the system's decisions based on the input signals and generates output signals which operate accuators to carry out the decisions.

The fourth general function is that of controlling. The controller is the brain of the control system. It receives the inputs from the sensors and external devices and performs mathematical calculations and logical comparisons to decide what must be done next. It then generates the correct output signals to the actuators that will carry out the decision. The controller usually is made up of some combination of amplifiers, filters, and digital circuits, including logic devices and microprocessors. It is not the intention of this book to address the design of the controller itself, but rather how certain types of controllers can be used to automate processes. If the reader is interested in knowing more about microprocessors, we recommend another of the books in the Understanding Series™: D. L. Cannon and G. Luecke, *Understanding Microprocessors*, Texas Instruments Incorporated, 1979.

Example of Component Use

"Automatic" control of devices has been around a long time. An early example of a control process illustrates the components included in a control system. A Roman emperor devised a scheme to have his temple doors open "by magic" as he approached. The system is shown in *Figure 2-7*. As he approached the temple, someone would build a fire on the altar. The heated air in the tank under the altar would expand and displace water from the hidden sealed tank into the bucket. As the bucket became heavier than the counterweight, it would drop slowly and open the doors. When the fire was quenched as the emperor left, the air cooled and the water was siphoned back to the sealed tank. As the bucket became lighter, the counterweight closed the doors.

**Figure 2-7.
Emperor's Control
System**

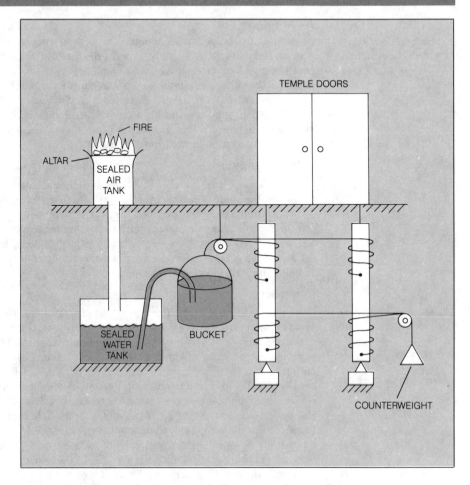

In this example, the sensor is the person who sees the emperor approaching and builds the fire. The fire is the signal conditioner that changes the visual sighting to an appropriate control signal. The heated air is the input signal to the controller which consists of the sealed tank, water and bucket. The pulley, rope, poles, and counterweight are the actuators.

TYPES OF SIGNALS IN CONTROL SYSTEMS

For the purposes of this book, all control systems can be classified into one of two types: analog or digital. If the signals into and out of the controller itself are continuous signals, then the control system is called an analog control system. On the other hand, if the signals into and out of the controller itself can be only in two different states identified by two voltage levels or two current levels, the system is called a digital control system. Let's investigate these definitions a little further.

Analog Signals

The level or amplitude of analog signals may vary continuously with time. That is, the amplitude of the signal may assume any of an infinite number of values between some maximum and minimum limits. *Figure 2-8* shows a common analog signal called a sinusoid or sine wave because its waveshape or waveform is described by the mathematical sine function. It is called a periodic signal because the waveform repeats every T seconds. The time interval in seconds, T, is the period of the signal. If 1 is divided by the time interval, then the frequency of the signal is obtained and is designated by the letter f. In like fashion, if the frequency of a signal is known, dividing it into 1 will give the period. Therefore,

$$f = \frac{1}{T} \text{ and } T = \frac{1}{f}$$

The frequency indicates how many times the signal repeats in one second or how many cycles there are per second (cps). The unit of frequency is hertz (Hz) and one Hz equals one cps. The alternating current electrical signal used for power in homes in the U.S. is a sinusoid with a frequency of 60 Hz. It is commonly called 60 cycle ac.

An analog signal varies in a continuous wave-like fashion over a given period of time, and has an infinite number of values between its maximum and minimum limits. Usually such signals vary repetitively at some frequency.

Figure 2-8.
A Periodic Continuous Signal — The Sinusoid

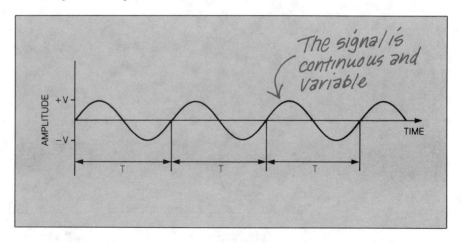

The signal is continuous and variable

Not all continuous signals are periodic and not all periodic signals are sinusoids, but all can be described in terms of sinusoids. This is very important because we express the frequency content of a signal in terms of sinusoids. A filter is a device that is designed to reject some frequencies in a signal while letting other signals pass unaffected. For example, radio and televison receivers contain a tunable filter that allows the signal of the selected station to pass while rejecting all others.

Most sensors generate analog signals and are easily attached to an analog controller (through signal conditioners if necessary). The frequency of the sensor signal may cover a band of frequencies from less than one hertz to thousands of hertz; however, in manufacturing systems, the sensor signal normally varies at a frequency less than a few hertz. A radio wave, on the other hand, has frequencies measured in hundreds of thousands to millions of hertz.

Digital Signals

The digital signals used in control systems have one of two possible levels; therefore, they are binary signals. We can represent this signal in numerical form by assigning one of the levels a value of "1" and the other a value of "0", the two values in the binary (base 2) system. Digital electronic circuits operate on voltage levels, but the voltage levels represented by a 0 or a 1 in one system may be different from the voltage levels represented by a 0 or a 1 in another system. For example, one system may use a level of +0.4 volt for a 0 and a level of +2.4 volts for a 1. Another system may use -0.4 volt for a 0 and +0.4 volt for a 1. However, this is important only to a circuit designer or troubleshooter, so the numerical (binary) representation is more nearly universal and easier to use when discussing digital signals and the information they represent. Therefore, when plotting digital waveforms when the actual voltage levels are not important, the binary 0 is represented at the 0 level of the Y axis and the binary 1 is assigned some arbitrary positive level as shown in *Figure 2-9*.

**Figure 2-9.
A Digital Signal and Its
Binary Representation**

Digital signals operate in a binary mode (only two signal levels are used, 1 and 0). A grouping of bits represents a code. A common grouping is a byte which is 8 bits.

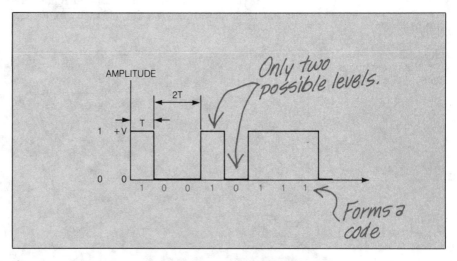

Like analog signals, digital signals have a period and frequency, but the period is measured differently. As shown in *Figure 2-9*, the period, T, is the time the level is at that of one binary digit (bit). This time varies from one system to another, but it must be within specified limits for any particular system. The frequency, usually called the bit rate, is the reciprocal of the period (1 divided by T) and is measured in bits per second (bps).

Since one bit can represent only two conditons such as ON/OFF, A/B, etc., when many different conditions are required, several bits are used as a group to form a code. Each combination of bits that can be formed from the group represents one of the different conditions. A common grouping used in digital systems has 8 bits and is called a byte. Such an 8-bit group is shown in *Figure 2-9*. This particular combination of bits can be represented by writing the 0 and 1 values as the binary number 10010111. An 8-bit code can have 256 different combinations to represent 256 different conditions.

If the sensors and actuators in a control system generate or use analog signals, but the controller needs digital signals, the signal conditioning required between the sensor and the controller is called an analog-to-digital converter and that required between the controller and the actuator is called a digital-to-analog converter. As the name implies, these units cover one signal form to the other; they will be explained further in Chapter 4.

STRUCTURE OF CONTROL SYSTEMS

So far in this chapter, the kinds of processes that may need to be controlled (continuous, batch, discrete parts) and the types of systems available to perform the control function (analog or digital) have been discussed. Now, the methods for controlling a system and the characteristics of system performance associated with these methods will be examined.

The block diagram shown in *Figure 2-6* describes a general control system. This means that all elements likely to be found in a control system are shown, but the elements themselves are not described. The system described is called a closed-loop system. Most systems that will be discussed will be closed-loop systems. But first, to better understand closed-loop systems, let's look at an open-loop system. It is a less complex system that is satisfactory for many applications.

Open-Loop System

In an open-loop control system no information is fed back from the output. As a result, input settings determine the output and, depending on the system response, can result in large output errors.

Figure 2-10 depicts a typical open-loop control system. A process is controlled by inputting to the controller the conditions believed necessary to achieve the desired result and accepting whatever output results. Cooking a roast in an oven with timer controls while we are away from home is an example of an open-loop system. We place the roast in the oven, set the cooking time and temperature, then remove the roast when we come home. The inputs to the control system are the time and temperature setting, the controller is the timer and thermostat which determines when to turn the oven on and off, the actuators are the switches that control the application of electricity to the heating element (or valves that control the flow of fuel to a burner), the process is the cooking, and the output is the roast. In this simple controller, no signal conditioning is required.

Figure 2-10.
An Open-Loop Control System

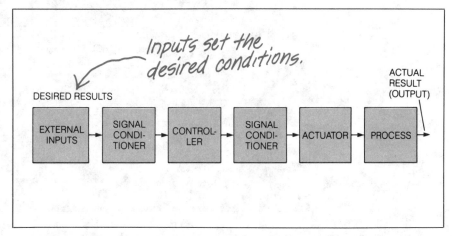

Compare *Figure 2-10* with *Figure 2-6* and note that the process sensors and the feedback loop are missing from *Figure 2-10* because it is an open-loop system. In an open-loop system, no information is fed back to the controller to determine if the desired result is being achieved. Large errors can develop in the output because of this. In the oven example, the roast may be overcooked or undercooked.

Closed-Loop System

In a closed-loop system, adjustments are made continuously by the control system until there is little difference between the actual output and the desired output.

A closed-loop system is a system that measures the actual output of the process and compares it to the desired output. Adjustments are made by the control system until the difference between the desired and actual output is as small as required. *Figure 2-11* shows a typical closed-loop system. The signal conditioning and actuators are not shown for simplicity and can be assumed to be part of the controller block if they are necessary.

Notice that the actual output is sensed and fed back (hence the name "feedback control") to be subtracted from the input that indicates what output is desired. If a difference occurs, a signal to the controller causes it to take action to change the actual output until the difference is zero. In the earlier example of cooking the roast, if a meat thermometer (a sensor) were placed in the roast, and if the oven were controlled to the set temperature, and if the roast were cooked until the meat thermometer indicated 140°F as measured by the control system, and then the oven were turned off, the control system would be a closed-loop system. The external inputs would be the start time, the oven temperature and the desired meat temperature. The input to the controller would be the difference between the desired and actual meat temperature. When that difference reached zero, the controller would turn off the oven. Actually, the controller would anticipate that some additional cooking would occur after the oven was turned off and would stop heating the oven when some difference, say 2°F, still remained. Even this example implies that different levels of closed-loop control are possible.

**Figure 2-11.
A Typical Closed-Loop
System**

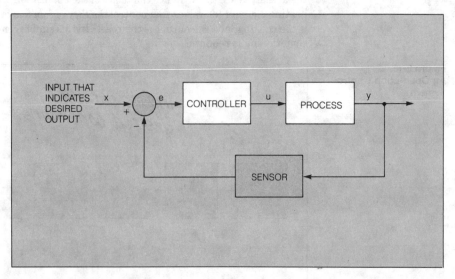

The automated filling system in *Figure 2-12a* is another example of a closed-loop process. An empty container is moved into position by the input conveyor system, filling begins, and the weight of the container and contents is monitored. When the actual weight equals the desired weight, filling is halted, the filled container is moved to the output conveyor, the next empty container is moved into place, and the process is repeated.

The control system block diagram is shown in *Figure 2-12b*. A sensor attached to the scale weighing the container generates the voltage signal or digital code that represents the weight of the container and contents. This is subtracted from the voltage signal or digital code that has been input to represent the desired weight. As long as the difference is greater than zero, the controller keeps the valve open. When the difference becomes zero, the valve is closed. This type of system is particularly useful for filling when the weight of the material being placed in the container is not directly related to the volume of the material. (Cereal products and potato chips are good examples of this.)

An automated filling system is an excellent example of a closed-loop system.

Figure 2-12.
Container Filling System

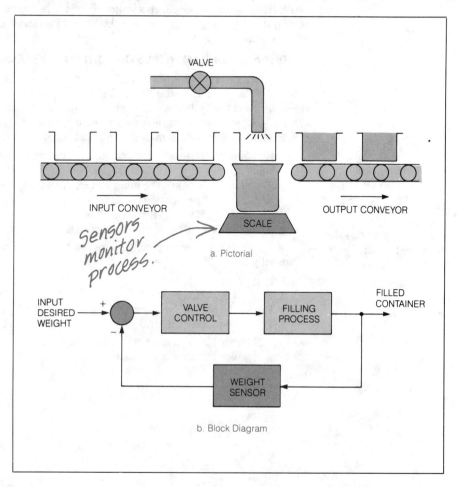

a. Pictorial

Sensors monitor process.

b. Block Diagram

Even a straight forward closed-loop system like an automated filling system requires several vital functions to be controlled and adjusted simultaneously.

Several questions may come to your mind when thinking about this filling system because several things must happen and be under control to accomplish this relatively simple task. For example, the conveyor belts need to be controlled to start, move and stop to position the container. The control system needs a sensor input that tells when the container is positioned for filling. Some time must be allowed for the scale to settle before reading the weight. Some means must be provided to move the container from the input conveyor to the scale and from the scale to the output conveyor. All of these problems can be handled easily by the controllers to be introduced in this book. The control system block diagram will be a little more complex than the ones shown so far, but no harder to understand. Your goal right now, however, should be to understand what is involved in the basic open-loop and closed-loop systems.

Use of closed-loop control systems are not without some disadvantages. In addition to being more complex and more expensive than open-loop systems, they have the potential to cause the output to oscillate with ever increasing amplitude. Such a system is said to be unstable and will usually destroy itself if allowed to continue in this mode. This and other characteristics will be discussed next.

THEORY AND FUNDAMENTALS OF CONTROL SYSTEMS

Refer again to *Figure 2-11*. The system performance will be discussed in terms of the symbols on the figure. The process output variable y is to match the input x and the system will attempt to accomplish this by calculating the error, $e = (x - y)$, and generating a process input signal, u, based on the value of e. System performance is measured against standard (criteria) such as:

The quality of an automated system's performance is determined by measuring the systems stability, sensitivity, responsiveness and other such parameters affecting the desired output.

1. How rapidly the process output variable y responds to a change in input x,
2. The error, e, between the input signal, x, and the process output variable, y, after a period of time,
3. The stability of the system, and
4. The sensitivity of the system to disturbances.

For process control of continuous and batch systems, the controller will usually be one of four types:

1. Proportional,
2. Proportional plus integral,
3. Proportional plus derivative, or
4. Proportional plus integral plus derivative (three mode).

Number 1 is the least complex and least accurate while number 4 is the most complex and most accurate.

The Transient Response

When the input x to a control system is changed suddenly, the output y with respect to time will respond to the change in a certain way called the transient response. This response will take on one of three general shapes as shown in *Figure 2-13*. The input is assumed to have changed suddenly from 0 to 1 and the output attempts to achieve the value of one as time passes.

The first type of response is called the underdamped response. The output y overshoots the value dictated by the input, then undershoots the value, and finally settles to a value close to the input value. This response is said to have an oscillating or ringing effect. A second possible response is the overdamped response where the output y does not ever overshoot the value dictated by the input, but takes a relatively long time to reach its final value. The third possible response is the critically damped response where the output y reaches its final value in the minimum possible time without overshooting the value dictated by the input x.

Both mechanical and electrical control systems have response time characteristics. The mechanical damping of an automobile suspension system is a familiar illustration of these characteristics. A "hard riding" sports car may be overdamped while a car with weak shock absorbers is definitely underdamped. The best ride is achieved by designing the car to be slightly underdamped.

A given process may allow oscillations about the final value with no adverse affects. Others may require that the output never exceed the input. The controller must be chosen or adjusted to accomplish whatever is required.

The sudden adjustment of the system will result in one of three types of transient responses: underdamped (swings under and over the desired level until the results are achieved); overdamped (no overshoot, slow rise to desired level); and critically damped (rapid rise to desired level without going over).

**Figure 2-13.
Types of Transient
Response**

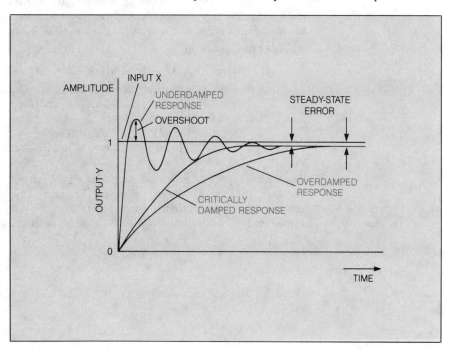

Steady-State Error

After the transient period, a final steady-state value of the output y will be reached for a given input x. The difference between the final steady-state output and the value dictated by the input is called the steady-state error. The steady-state error for an input indicating a position is shown in *Figure 2-13*.

Stability

If a system becomes un-stable, it has the possibility of doing extensive damage to system parts.

If for a given input or disturbance the transient response of the system is such that a steady-state output is achieved like that shown in the *Figure 2-13*, the system is said to be stable. If the system is unstable, the output will continue to increase without limit until the system destroys itself unless safety circuits are included to shut down the system. For example, on November 7, 1940, the first bridge across the Tocoma narrows at Puget Sound, Washington, was subjected to a wind gust (disturbance) and began to oscillate with increasing amplitude until it broke apart.

Sensitivity

The sensitivity of a system is the ratio of the percent change in output y to the percent change in inputs to the system. These inputs may be the normal ones or they may be unwanted disturbances. Process parameters change due to aging, environment, improper calibration, etc. Closed-loop systems are much less sensitive to these changes than are open-loop systems because the closed-loop systems monitor the output and can compensate for these changes. If highly accurate open-loop systems are needed (low steady-state error), the components of the system must be selected very carefully and the cost of the system skyrockets.

Types of Systems

The three types of systems which will be discussed are:

Type 0: A constant input signal x results in a constant value (constant position) for the controlled output variable, y.

Type 1: A constant input signal x results in a constant rate of change (constant velocity) for the output variable, y.

Type 2: A constant input signal x results in a constant acceleration for the output variable, y.

The system type is determined by considering the controller and the process together. The example shown in *Figure 2-12* is a type 1 system because a constant input x results in a constant flow (velocity) of material to fill the container. Each of the system types has a different steady-state response.

Type 0

The steady-state response of a type 0 system to a step input is shown in *Figure 2-14a*. If the system has a gain of K, then the steady-state error, e_{ss}, for a step (also called a position or setpoint) input of value A is:

$$e_{ss} = \frac{A}{1 + K}$$

The larger the value of K, the smaller the error, but large values of K may make the system unstable. If the input to a type 0 system is a velocity or an acceleration input, the output cannot follow it and the steady-state error increases with time and approaches a value of infinity.

In a type 0 system, a constant compensation value is made to respond to input results in a constant output that is different from the input by a constant error.

Figure 2-14.
Steady-State Error for
Different System Types

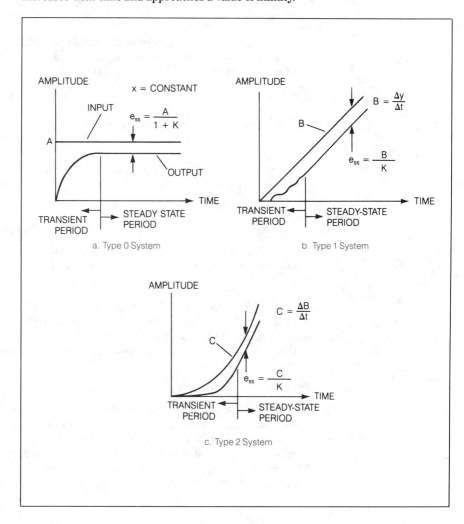

a. Type 0 System

b. Type 1 System

c. Type 2 System

Type 1

The steady-state error of a type 1 system to a step input is zero, which is the desired situation. The steady-state error of a type 1 system due to a ramp input of magnitude B is shown in *Figure 2-14b*. Again K is the system gain and the steady state error, e_{ss}, is:

$$e_{ss} = \frac{B}{K}$$

As before, increasing K decreases steady-state error. A type 1 system cannot follow an acceleration input and the steady-state error for an acceleration input approaches infinity as time approaches infinity.

Type 2

A type 2 system has zero steady-state error for both position and velocity inputs. If the input is an acceleration of value C, then the steady-state error, e_{ss}, is as shown in *Figure 2-14c* and is calculated by:

$$e_{ss} = \frac{C}{K}$$

Again, increasing K decreases steady-state error.

Now, let's look at the different kinds of controllers that can be used to improve system response.

Types of Controllers

Proportional Controller

A proportional controller simply adjusts the system gain, K. If the process has a gain K_p and the controller has a gain K_c, then the system gain is

$$K = K_c K_p$$

If the system is unstable for the gain K_p, we would choose a value for K_c that is less than one such that the system gain K results in a stable system. On the other hand, if stablity is not a problem, we can use a proportional controller and increase the gain until the steady-state error is decreased to the value desired.

Proportional Plus Integral Controller

A proportional plus integral (PI) controller (also called a lag compensator) adjusts the system gain the same as a proportional controller, but it also increases the type of the system by 1. Thus, if the process itself is a type 1 and a PI controller is added, the system is a type 2. As indicated in *Table 2-1*, increasing the type of the system increases the kinds of input that can be accepted by the system without generating unacceptable steady-state errors.

Type 1 systems have an output that changes at the rate of the input, but is different from the input by a constant error.

Type 2 systems have an output whose acceleration is the same as the input acceleration, but is different from the input by a constant error.

The Proportional Controller gain is increased until the steady state error is reduced to within acceptable limits. A PI controller would have the same gain but the system type would increase by 1.

**Table 2-1.
Steady-State Error as a
Function of System Type
and Input Type**

Input Type	System Type		
	0	1	2
position, A	$\dfrac{A}{1+K}$	0	0
velocity, B	Infinity	$\dfrac{B}{K}$	0
acceleration, C	Infinity	Infinity	$\dfrac{C}{K}$

Proportional Plus Derivative Controller

The proportional plus derivative (PD) controller (also called a lead compensator) permits alteration of the transient response of the system. The use of a properly designed PD controller can change an underdamped system to a critically or overdamped system or vice versa within constraints established by the process itself. As with the other proportional controllers, the gain can be changed to alter stability and steady-state error. The type system changes as with the PI controller; however, it is not used a great deal by itself.

Proportional Plus Integral Plus Derivative Controller

The Proportional Plus Integral Plus Derivative Controller combines all three types of controller operations. It has zero position, zero velocity, and constant acceleration steady state error to a signal input with constant acceleration.

The proportional plus integral plus derivative (PID) or three mode controller is a combination of all three controllers discussed above. It allows alteration of gain, system type and transient response in order to improve the process operation. Determining the system gain to keep the system stable and yet give the response desired becomes more difficult as the system complexity increases. In addition, usually as functions are added more hardware is required.

CONTROLLERS FOR DISCRETE PROCESSES

In discrete parts manufacturing, controlling the process usually means turning on or turning off various components of the process (drills, saws, feeders, motors, etc.) for a given time in sequence. Many of the inputs are generated by contact closures (or openings) from various switches and relays that are operated by a person, by the process, or by timers. Some inputs for process-controlled switches come from various sensors that measure temperature, flow, etc. Control programs for these on-off, start-stop processes are usually developed by using a ladder diagram. This diagram shows under what conditions each of the components of the process will be connected to the power source. These ladder diagrams and controllers will be covered in detail in Chapters 5, 6, and 8.

WHAT HAVE WE LEARNED?

1. There are three kinds of processes that may need control: continuous, batch, and discrete parts.
2. Analog or digital controllers may be used. Analog controllers interface easily to sensors and actuators, but are difficult to adapt if requirements change. Digital controllers require special circuitry to interface to sensors and actuators, but adapt easily to change.
3. All control systems are concerned with transient response, steady-state error, sensitivity to change, and stability.
4. A control system may be open-loop or closed-loop. Open-loop systems are less expensive, but closed-loop systems are less sensitive to change and disturbance and have a smaller steady-state error.
5. Four types of controllers for continuous and batch systems are:
    a. Proportional (increases stability),
    b. Proportional plus integral (increases stability, decreases steady-state error and increases the types of inputs allowed),
    c. Proportional plus derivative (adjusts transient response), and
    d. Three mode or proportional plus integral plus derivative (combines advantages of all the above).
6. For discrete parts processes, discrete component controllers or programmable controllers can be used.

Quiz for Chapter 2

1. A continuous process is:
  a. one that never shuts down.
  b. used only for simple tasks.
  c. self–contained in that raw materials enter the process and an identifiable product exits the process.
  d. used only with analog controllers.

2. A discrete parts process is:
  a. often encountered in manufacturing.
  b. the same as a continuous process except a different controller must be used.
  c. normally a repetitive series of operations.
  d. a and c.

3. Components of a control system may include:
  a. sensors.
  b. actuators.
  c. amplifiers.
  d. all of the above.

4. Analog and digital controllers:
  a. accept the same input signals.
  b. cost the same to build.
  c. can be changed with equal ease.
  d. none of the above.

5. If the period of a sine wave is 0.001 second, the frequency of the sine wave is:
  a. 0.001 Hz.
  b. 1000 Hz.
  c. 10 Hz.
  d. 10,000 Hz.

6. If the transient response is underdamped:
  a. the value of the output will exceed the value of the input at times during the transient period.
  b. steady–state will be reached faster than if the system is overdamped.
  c. the percent overshoot is zero.
  d. a and b.

7. The steady–state error is:
  a. a function of the transient response.
  b. independent of the type of input.
  c. zero for all inputs to type 1 systems.
  d. decreased by increasing gain.

8. A proportional plus integral controller can:
  a. can compensate for lag.
  b. has no transient response.
  c. can increase the system type.
  d. a and c.

9. Stability:
  a. can be increased by increasing gain.
  b. can be increased by decreasing gain.
  c. is not important.
  d. none of the above.

10. Which of these controller types is the most accurate?
  a. proportional.
  b. proportional plus integral.
  c. proportional plus derivative
  d. three mode.

11. For a type 0 system, the steady-state:
  a. error for a step input is always zero.
  b. error for a velocity input is always zero.
  c. input for acceleration input is always non-zero.
  d. a and c.
  e. all of the above.

12. For a type 1 system, the steady-state:
  a. error for a step input is always zero.
  b. error for a velocity input is always zero.
  c. input for acceleration input is always non-zero.
  d. a and c.
  e. all of the above.

13. For a type 2 system, the steady-state:
    a. error for a step input is always zero.
    b. error for a velocity input is always zero.
    c. input for acceleration input is always non-zero.
    d. a and c.
    e. all of the above.

14. Which of the following devices could be part of a sensor?
    a. thermometer
    b. brake
    c. relay
    d. clutch

15. Which of the following devices could be part of an actuator?
    a. thermometer
    b. scale
    c. barometer
    d. clutch

16. If the transient response of a system is overdamped:
    a. the value of the output exceeds the value of the input at times during the transient period.
    b. the value of the output never exceeds the value of the input during the transient period.
    c. the value of the output never stabilizes.
    d. none of the above.

17. Compared to an open-loop system, a closed-loop control system is:
    a. more accurate.
    b. more complex.
    c. more stable.
    d. all of the above.

18. If a sine wave has a frequency of 500 Hz, the period in seconds of the sine wave is:
    a. 500
    b. 0.005
    c. 0.02
    d. 0.002

19. The general type of process found in industry is:
    a. continuous.
    b. discrete parts.
    c. batch.
    d. all of the above.

20. Digital signals:
    a. can have three possible levels.
    b. are assigned values of 1 or 0 depending on the voltage level.
    c. have a frequency measured in units of hertz.
    d. have a sinusoidal shape.

Basic Control System Hardware

ABOUT THIS CHAPTER

Industrial automation means or implies the practical application of concepts and ideas in order to improve the manufacture of a product, to make the product better, or, in many instances, to make the product at all. Automation systems have been designed using many different technologies. The two technologies most widely used in the past have been relay logic and pneumatic controls; however, the availability of electronic components, especially the microprocessor, has made the design of automation systems using electronic technology much more attractive.

In this chapter, the hardware used to build an electronic control system will be described. The hardware can be divided into three different parts: the controller, which supplies the decision-making capability, sensors which supply the input information to the controller, and the actuators which supply the means for a control to perform some form of mechanical action. The emphasis will be on the components used in electronic control applications.

AUTOMATION SYSTEM OVERVIEW

The main parts of an automated control system are: the input sensors, the controller and the output actuators. There is some form of operator and centralized computer interface, but they do not control the process directly.

Chapter 2 shows the concept of a control system accepting a sensor input, making a decision, supplying an output and, in a closed-loop system, comparing the output to what is desired. This provides a good overall view, but does not give much insight into how a practical system might actually appear. Since this chapter is concerned with the use of actual hardware, a better diagram for this purpose is the generalized hardware functional block diagram of a system shown in *Figure 3-1*.

As indicated by the dotted lines, the system consists of three main parts: the input sensors, the controller, and the output actuators. The controller block also includes any signal conditioning such as amplifiers, A/D or D/A converters required to translate the signals to a form acceptable to the electronic controller. In any practical system, there also will be some form of operator interface and, most likely, an interface to a higher level controller such as a main plant computer. Although these last two interfaces are vital to the overall process, they are there primarily to monitor the process or change the process parameters, not to control the process itself.

**Figure 3-1.
Control Hardware
Diagram**

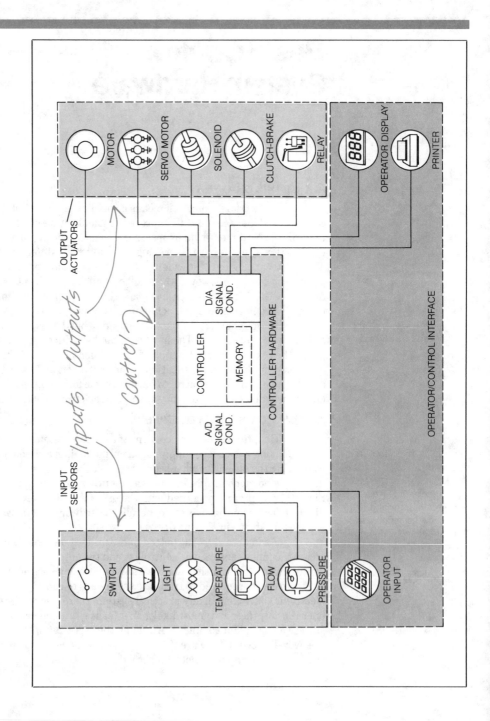

CONTROLLERS

The sequential and proportional controllers are the most common types of controllers and the older types are rapidly being replaced by low cost, reliable, electronic programmable processors.

As discussed in Chapter 2, a controller receives information about a process, makes a decision based on this information, performs some action on the process, and monitors the results. This basic concept has not changed with changes in technology, but several technologies have been developed to perform the controller functions over the years. Relay logic and pneumatic controls are two technologies that are still in widespread use. Relay logic control finds application in automation of sequential type machines while pneumatic control usually is found in automation of proportional control of a continuous process.

Both sequential and proportional process types can be handled by a relatively new product; programmable electronic controllers based on solid-state programmable processors. The availability of low-cost, reliable, solid-state electronic components and processors has made the electronic programmable controller the emerging leader in industrial control.

Electronic programmable controllers are "software programmable"; that is, the instructions are stored in a memory and the instructions can be changed easily and quickly. To change instructions in a relay or pneumatic controller requires physical wiring or tubing changes which often is difficult, expensive, and time consuming, if it is possible at all.

CONTROLLER TYPES

There are two types of controllers available to perform industrial control. The most basic is the sequence controller which is concerned with the ordering of a control process and deals with ON/OFF control signals that occur step-by-step or in repeated steps, or in a combination of the two, to accomplish a task. The second is the process controller which is concerned with the control of a continuously variable process such as might be encountered in a chemical manufacturing process.

Sequence Controller

A sequence controller is generally used in open loop systems in which sequential events are being controlled or discrete products are being manufactured.

A sequence controller is usually concerned with the manufacture of a discrete product or with the sequencing of a number of events in a plant. Basically it performs ON/OFF control functions in relation to time. It usually is an open loop type controller. The drum timer cycle control is a good example of an electromechanical sequential controller. It has been used for years to control home appliances such as the automatic clothes washer. A timer motor rotates one or more cams to cause different switches to open and/or close which in turn control the operation of the machine. The speed of the timer motor and the shape of the cams determine the amount of time allowed for each operation. The relation of the cams to each other determine the sequencing.

The drum timer can best be visualized as a cylinder with a matrix of decision-points as shown in *Figure 3-2a*. *Figure 3-2b* shows a table that represents the data used in a drum timer application. The left column gives the step number and step time interval. The remaining columns represent individual output points. A "1" or a "0" signifies whether or not power is applied to an output during each step increment. The output state plotted against the step increment becomes the "program" for the drum timer.

Instead of using a mechanical drum timer with cams and switches, an electronic programmable controller can do the operations much easier, and many more features can be added that the simple drum timer does not provide. For instance, the sequence of steps and each step time increment for each operation can be individually programmed and reprogrammed without any rewiring or repositioning of cams. This detail will be expanded further as applications are discussed in later chapters. For now, let's concentrate further on the control system itself.

Process Controller

A process controller is concerned with the control of a continuously variable process that requires constant monitoring and continual updating of inputs and outputs in order to maintain a very precise setpoint. Instead of discrete ON-OFF levels, the process controller provides an output which can vary continuously from full on to full off. In many applications, the process controller must be able to perform mathematical operations in order to provide the control.

In order to understand the operation of a process controller, let's look at an example. Both solutions in this example have the effect of controlling the parameter, but the method and the degree of control are quite different.

The best choice for a continuous process controller is a proportional controller or a limit cycle controller.

The first solution, shown in *Figure 3-3a*, provides a means of maintaining temperature control using limit cycle control. It is the method commonly used in a home heating system. The operator sets the desired nominal temperature. This is the setpoint. The thermostat makes the decision to turn the furnace ON or OFF depending on whether the temperature is below or above the set point. Typically a deadband of 2 to 4°C is used to prevent the furnace from cycling on and off every few seconds. This increases the efficiency of the system. Even if the deadband is zero, there is a lag between the sensor input and the furnace actuation that results in some overshoot or undershoot of the desired temperature. Using the limit cycle approach, the temperature will oscillate between the lower and upper limits as shown in *Figure 3-3a*.

The second solution uses proportional control. As shown in *Figure 3-3b*, a process controller is used that can provide a continuously variable signal to a fuel flow control valve that can be adjusted continuously to provide a continuously variable amount of heat from the furnace. The system response is shown in *Figure 3-3b*. In this case, the amount of heat produced can be adjusted to just balance the amount of heat lost and the temperature can be maintained very close to the setpoint. The response shown in *Figure 3-3b* is typical of the system response to a disturbance or setpoint change.

**Figure 3-2.
Drum Timer**

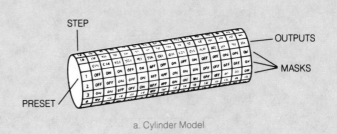

a. Cylinder Model

		C	C	Y	Y	Y	Y	C	C	C	C	C	Y	C	Y	C
DRUM 01																
PRESET = 02		0	1	0	0	0	0	0	0	0	1	0	0	1	0	1
SCN/CNT = 32000		2	4	1	3	4	5	1	3	4	5	5	8	2	2	3
STEP	CNT/STP*															
1	00002	0	1	1	0	1	0	0	1	0	1	0	0	1	1	1
2	00001	0	0	1	0	1	0	0	1	1	1	0	0	0	0	1
3	02000	1	1	0	0	1	0	0	1	1	1	0	0	0	0	1
4	00000	0	1	0	0	1	0	0	1	0	1	0	0	0	1	1
5	00202	0	1	1	0	1	0	0	1	0	1	0	0	0	0	0
6	10000	0	1	1	0	1	0	0	1	1	0	0	0	1	1	0
7	00003	0	1	1	0	1	0	0	0	1	0	0	0	0	0	0
8	00010	0	0	1	0	1	0	0	1	1	1	0	0	1	0	0
9	00001	0	1	0	0	1	0	0	0	1	1	0	0	0	1	0
10	02000	1	1	0	0	1	0	0	1	1	1	0	0	1	0	0
11	01230	0	1	1	0	1	0	0	1	1	1	0	0	0	0	1
12	00300	0	0	1	0	1	0	0	1	1	1	0	0	0	0	1
13	00001	1	1	1	1	1	0	0	0	0	0	1	1	1	0	1
14	00001	0	1	1	0	1	0	0	1	1	1	0	0	1	0	0
15	00002	0	1	1	0	1	0	0	1	0	0	1	1	0	0	0
16	00002	0	1	1	0	1	0	0	1	0	0	1	1	0	0	0

*Counts/step indicates length of time outputs are in the prescribed state for each step.
1—indicates output will be ON during step.
0—indicates output will be OFF.

b. Programming Form

**Figure 3-3.
Limit Cycle Versus
Proportional Control**

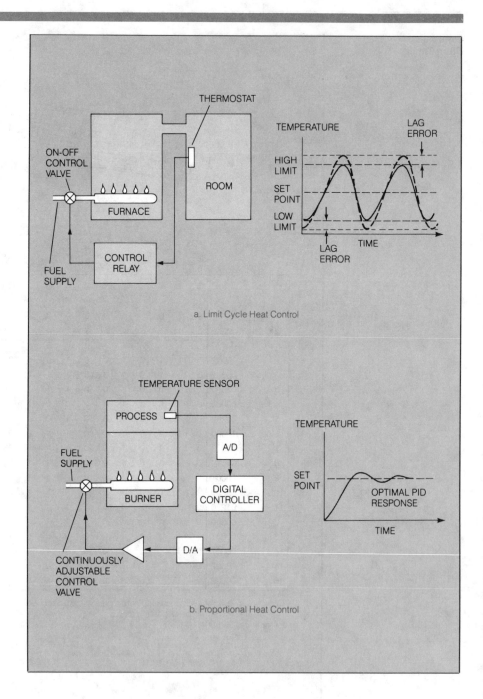

a. Limit Cycle Heat Control

b. Proportional Heat Control

SENSORS

The ability to control a process or machine is dependent first of all on the ability to sense what is happening. This has been and continues to be a major obstacle in the automation of tasks now performed by humans, especially those involving the eyes (vision).

Sensors should have the ability to measure accurately and reliably the physical parameters involved in the process.

The ideal sensor would be small in size, durable and reliable, and have infinite resolution and accuracy. Its output would not drift due to temperature or any other environmental factor and, of course, it would be easy to make and low in cost. Despite the fact that many sensors are available today, there continues to be a need for better and more accurate sensors, especially as the availability of low-cost reliable controllers makes the automation of more and more tasks feasible.

The basic sensor types are related to the physical properties needed to describe the world around us. The ability to measure these properties accurately is a necessary requirement in automating a process. There are many varieties of input sensors, from very simple switch contacts to very elaborate atomic particle detectors, all of which find applications in various automation systems.

PROCESS CONTROL SENSORS

Continuous process control usually is concerned with maintaining a number of different parameters within some previously determined limits. Three common parameters are temperature, pressure, and flow.

Temperature Sensor

Temperature is important in many process control applications. Often a reaction will take place only at a set temperature or when a certain amount of heat is added. The efficiency of some processes is affected by the temperature at which they are run. Three temperature sensors in current use are the thermocouple, the thermistor, and the RTD. An example type of each of these is shown in *Figure 3-4*.

Thermocouple

A thermocouple is a temperature sensor based on the principle that two joined, dissimilar metals generate a voltage when heated.

In the early 1800's, Thomas Seebeck discovered that a voltage was generated in a circuit consisting of two junctions of two dissimilar metals. When one of the two junctions was kept at a constant temperature, the change in voltage across the circuit became a function of the temperature of the other junction. This is the basis of the thermocouple temperature sensor. An illustration of the Seebeck voltage generation is shown in *Figure 3-5a*.

**Figure 3-4.
Temperature Sensors**

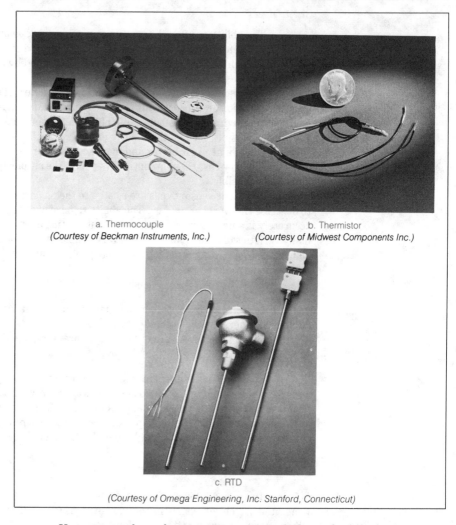

a. Thermocouple
(Courtesy of Beckman Instruments, Inc.)

b. Thermistor
(Courtesy of Midwest Components Inc.)

c. RTD
(Courtesy of Omega Engineering, Inc. Stanford, Connecticut)

Very accurate low voltage measurements must be made since the thermal voltage is only a few millivolts (mV) per degree C as indicated in *Figure 3-5b*. A typical measuring circuit is shown in *Figure 3-5c*. Despite this need, thermocouples are probably the most widely used temperature measurement device in their useful range because they are inherently accurate, inexpensive, and may be coupled easily to control equipment through signal conditioning amplifiers.

Thermocouples can be used over a temperature range of $-250°C$ to $2000°C$. Different combinations of metal cover different portions of this range. Theoretically, the upper temperature is limited only by the melting point of the materials; but practically, problems with brittleness and oxidation usually limit their use to temperatures well below their melting points.

Figure 3-5.
Thermocouple

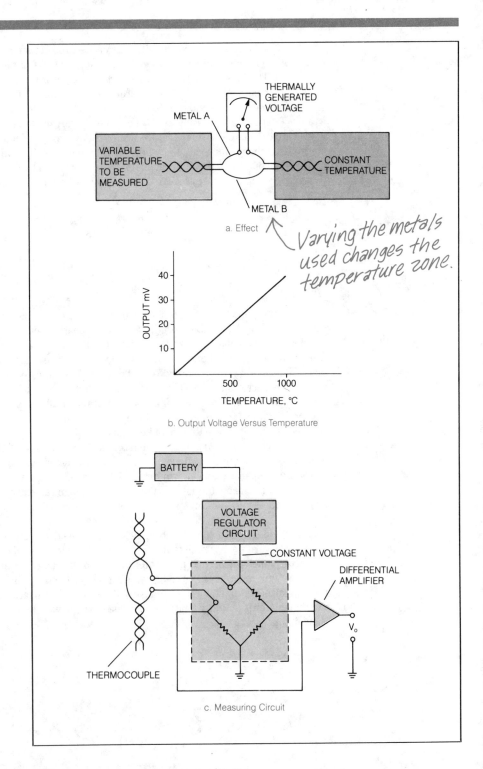

a. Effect

Varying the metals used changes the temperature zone.

b. Output Voltage Versus Temperature

c. Measuring Circuit

The electrical output in millivolts of one type of thermocouple versus temperature is shown in *Figure 3-5b* for the range of 0°C to 1,000°C (1 mV is 0.001 volt). The known relation for this particular sensor is that the output voltage is linear with respect to temperature and changes 0.04 mV for each °C change in temperature and that 0.0 mV = 0°C. This relation would be stored in a table of data in an electronic controller's memory. A program would be written to calculate the temperature when it is required and it also would be stored in memory. For example, when the controller received an electrical input signal of 32 mV, it would calculate the temperature by following the program through steps that solve the following equation:

$$\text{Temp }°C = \frac{\text{Input mV}}{0.04 \text{ mV/}°C}$$

$$= \frac{32}{0.04}$$

$$= 800°C$$

Thermistor

A thermistor is a heat sensor that measures the temperature by the change in a semiconductor's resistivity.

A thermistor is made of semiconductor materials such as nickel or cobalt oxides which exhibit a predictable and repeatable change in resistance as temperature is changed. As the material is heated, more electrons break free from the covalent bonds and this action reduces the resistance of the material as illustrated in *Figure 3-6a*. The change is indicated in *Figure 3-6b* and it is characteristic of NTC (negative temperature coefficient) material. In some cases, the resistance change can amount to 5 to 10 percent change per degree C, which for a 10,000 ohm element is a 500 to 1,000 ohm change. Due to this sensitivity, it is possible to make measurements to within 0.05°C or better if sensitive circuits are used. A thermistor can be used up to 1,000°C if special solderless packaging is used; however, the relative sensitivity decreases with increased temperature as shown by the flattening of the curve.

Since the output is an exponential change in resistance as shown in *Figure 3-6b*, it must be converted into a voltage or current change for use by a controller. A simple way to do this is to have the thermistor act as part of a voltage divider as shown in *Figure 3-6c*. The voltage drop across the thermistor varies as its resistance changes. This voltage is fed to an A/D converter which converts it into a digital code for use by the controller.

Care must be taken to be certain that the current through the thermistor is not high enough to cause self-heating. Self-heating means that the thermistor is actually heating itself due to the power dissipated in the thermistor.

**Figure 3-6.
Thermistor**

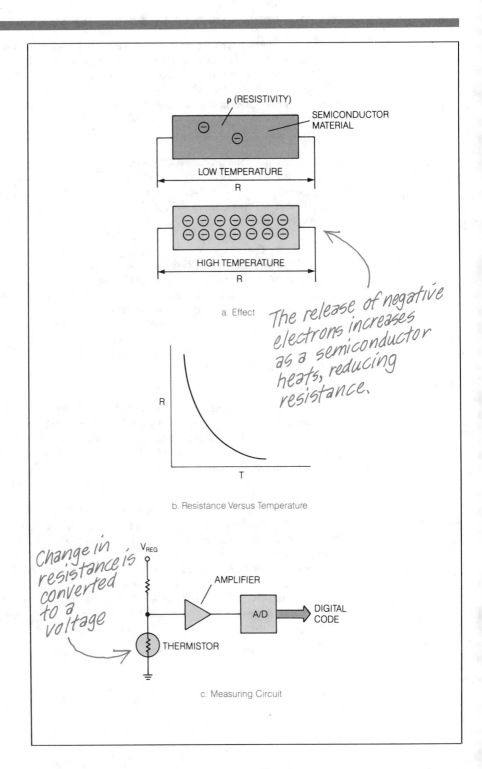

ρ (RESISTIVITY)

SEMICONDUCTOR
MATERIAL

LOW TEMPERATURE
R

HIGH TEMPERATURE
R

a. Effect

The release of negative electrons increases as a semiconductor heats, reducing resistance.

R

T

b. Resistance Versus Temperature

Change in resistance is converted to a voltage

V_{REG}

AMPLIFIER

A/D

DIGITAL
CODE

THERMISTOR

c. Measuring Circuit

Thermistor sensors are used when the range of temperature to be measured is small and when the resolution required is high. The disadvantage of the thermistor for industrial control is that its output is not linear with respect to temperature; therefore, some signal conditioning or special computer processing may be required. Thermistors have the advantages of being inexpensive and physically small.

RTD

Resistive temperature devices (RTD's) operate on the principle that the resistivity of a metal changes when heated. It has a linear change of resistivity with temperature.

One of the problems with a thermistor is that its resistance value can drift with age and must be recalibrated to maintain accuracy. A resistive temperature device (RTD) does not have this problem. Its operation is based on the fact that a metal when heated exhibits a change in resistance. This change is called its temperature coefficient of resistance. The RTD consists of a metal in pure form which has a very predictable and repeatable temperature coefficient. The most commonly used types are nickel, copper, and platinum, but platinum is used more because it can be obtained easily in pure form, does not oxidize easily, and has a fairly high temperature coefficient. Copper, on the other hand, is the least desirable, since it oxidizes easily, has a low melting point, and a very low temperature coefficient.

The resistance change is used as a controller input in a way similar to the thermistor circuit of *Figure 3-6c*. The RTD, however, has a linear change in resistance with temperature so that its output can be coupled directly to the A/D converter.

Pressure Sensors

The need to measure pressure in process control is a common requirement. Several devices are available to measure pressure, but we will discuss only the three most commonly used. None of the pressure sensors discussed provide an electrical signal directly, therefore, some type of conversion method must be used.

Aneroid Barometer

In an aneroid barameter, changes in pressure are converted to electrical signals by a moving diaphragm stretched over a sealed vacuum chamber.

An aneroid barometer is constructed by forming a chamber from rigid material except for one end which is a thin plate that can flex easily. Air is evacuated from the chamber and the chamber is sealed. The thin plate, called a diaphragm, deflects according to the pressure exerted on it. Some means of converting this deflection into an electrical signal is required to utilize it in a control system. One method is to mechanically link the diaphragm to the wiper of a potentiometer which has current flowing through it to produce a voltage drop as shown in *Figure 3-7*. When the diaphragm is deflected by a pressure change, it moves the wiper and changes the output voltage in relation to pressure.

**Figure 3-7.
Aneroid Barometer with
Potentiometer**

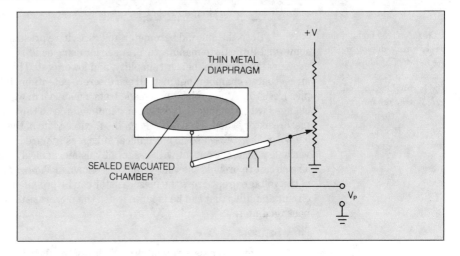

Capacitive Pressure Sensor

The plates on each side of
an evacuated chamber
form the capacitive pres-
sure sensor. Changes in
pressure deflect the plates
and change the
capacitance.

Another sensor that uses an evacuated chamber, but has two
diaphragms which act as the plates of a capacitor is shown in *Figure 3-8*. The
plates are insulated from the chamber by a non-conducting dielectric material.
As the plates deflect with pressure variations, the capacitance changes
because the distance between the plates varies. The capacitance can be
measured by providing an ac signal of known amplitude and frequency to the
capacitive sensor. As the capacitance changes, the output voltage changes in
proportion. A demodulator and phase detector supply an output voltage
proportional to pressure.

**Figure 3-8.
Capacitive Pressure
Sensor**

*As the plates
change with
pressure, the
capacitance
changes.*

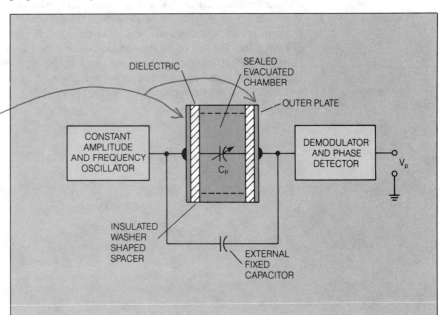

Solid-State Strain Gauge Pressure Sensor

A piezoresistor bonded directly onto a diaphram is the heart of a solid-state strain guage pressure sensor. It provides smaller size, better reliability and lower cost, than mechanical sensors.

Both the aneroid barometer and capacitive pressure sensors are somewhat difficult to manufacture. A sensor using solid-state technology provides smaller size, better reliability, and lower cost. This sensor also uses an evacuated chamber, but instead of a discrete potentiometer, it uses a piezoresistor. A piezoresistor is made from a material in which the resistance changes with a change in twisting or bending force on the material. The piezoresistor is deposited directly on the flexible plate of the sealed chamber as shown in *Figure 3-9a* so no mechanical linkage is required. Thus, it supplies a variable resistance output with essentially no mechanical linkage mechanism. The circuit to provide the voltage input is shown in *Figure 3-9b*. Amplification of the voltage drop across the piezoresistor can be provided by an integrated circuit amplifier that can be mounted with the piezoresistor to make a packaged unit.

Flow Sensors

Many processes require knowledge of the flow rate and mass flow of a material, usually a liquid or gas. Two flow measurement methods are used: One is based on the Bernoulli principle, the other uses a turbine.

Venturi

The difference in pressures at the point of constriction and upstream from the constriction is the basis on which venturi flow sensors operate.

One method of monitoring flow is to use a venturi. A venturi is a tube which has a constriction in it and its operation is based on the Bernoulli principle. The Bernoulli principle states that the pressure, velocity, and kinetic energy of a closed system is a constant anywhere in that system. Thus, the amount of material flowing through any cross-section of the tube at any given time is a constant. Since the velocity at the constriction must be higher to pass the same amount of material, the pressure must be lower according to the Bernoulli principle. Thus, the difference in the pressure at the constriction from the pressure upstream is proportional to the velocity of the material. Further, since the size of the tube and the constriction are known, the actual mass flow of the material can be determined. By placing pressure sensors at the constriction and upstream as shown in *Figure 3-10*, the controller can calculate the flow rate and mass flow of the material.

Figure 3-9.
Solid-State Strain Gauge
Pressure Sensor
(Source: W. B. Ribbens and
N. P. Mansour
Understanding Automotive
Electronics, *Texas Instruments*
Incorporated, Copyright ©
1982)

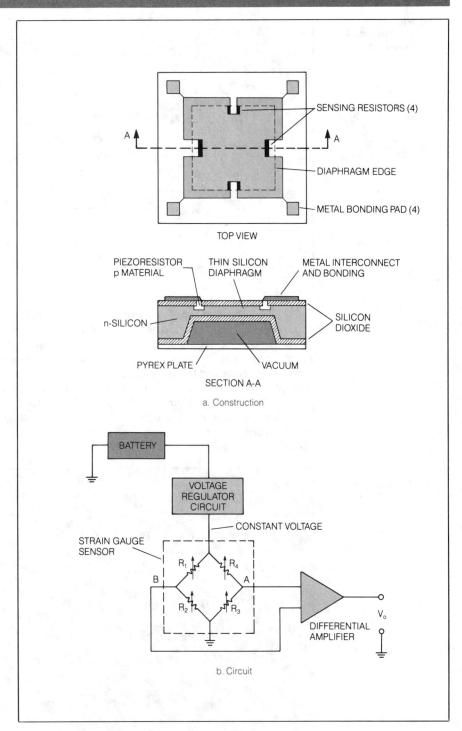

**Figure 3-10.
Venturi Constriction for
Flow Measurement**

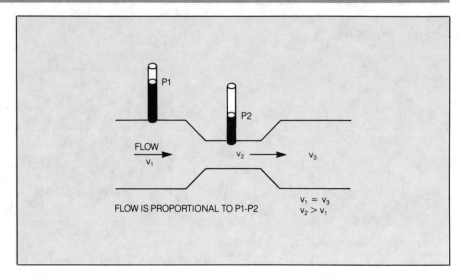

Turbine

Flow sensors must be
carefully chosen for their
specific application due to
the number of variables
that may affect their
accuracy.

Another method of providing a flow indication is with the use of a turbine. In this case, an impeller is placed in a tube through which the material flows. The turbine output shaft drives a generator which provides a voltage output proportional to speed. Since the output voltage versus speed of the turbine-generator and the cross-section of the tube are known quantities, the flow rate can be determined. A typical arrangement is shown in *Figure 3-11*.

A common problem with all types of flow meters is that accuracy is dependent on a number of variables that may not be known. Care must be taken to insure that the range and linearity are calibrated adequately for the specific application.

**Figure 3-11.
Turbine-Generator for
Flow Measurement**

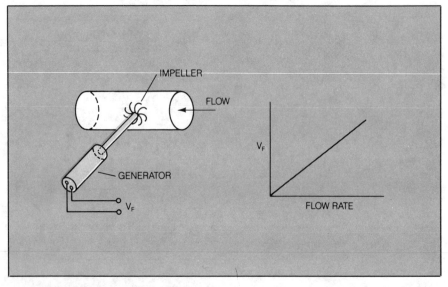

DISCRETE SENSORS

Sequential controllers may not require continuous inputs from many of its sensors, but rather ON/OFF indications only. Examples are: The limit of a pressure or temperature, flow over a maximum amount, and the position of an object. Position is one of the most important because often the actual position of a lever arm, motor shaft or other part must be known in order to perform an operation. Even in process control, the position of a control is often required knowledge for correct operation. Position sensors also can be used to count objects passing by or to count revolutions of a shaft.

Position Sensors

One of the simplest position sensors, but one which still is used widely today, is the mechanically activated limit switch. The closing or opening of its contacts can indicate position, but this type switch requires physical contact with something to actuate it. This is satisfactory for some applications, but for others, physical contact is not desirable or even permitted. One position sensor that does not require physical contact is the light sensor switch.

Light Sensor Switch

Position sensors using light are of two types. One interrupts a direct light beam another a reflected light beam.

There are many styles of light sensor switch that use photoelectric techniques to sense position. *Figures 3-12a and 3-12b* show two styles. One must have its light beam directly interrupted and the other uses a reflection of its light beam, but the basic operation is the same. A visible light or infrared source, often a light-emitting diode (LED), emits light. A detector senses the light, either directly or reflected, and produces an output. The detector is a photosensitive transistor which turns on when light energy is directed onto it. The result is the same as if a forward-biasing current were supplied to the transistor's base junction. A schematic of the LED emitter and phototransistor detector is shown in *Figure 3-12c*. By connecting the collector of the transistor to a logic level voltage through a resistor, the voltage at the collector will be high in the dark state and low in the light state. This signal can be coupled directly to the digital logic controller so conversion is not necessary. *Figure 3-13* shows the application of a interrupted light sensor to measure the revolutions per minute of a shaft.

**Figure 3-12.
Light Sensor Switches**

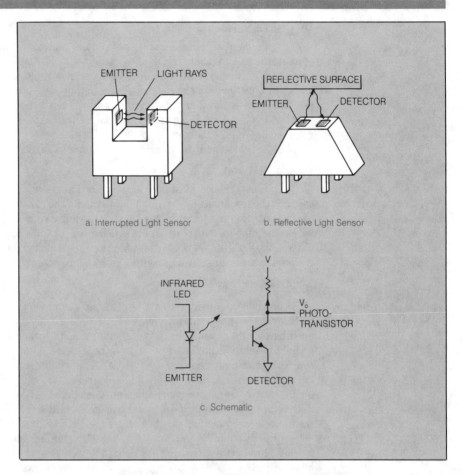

EMITTER   LIGHT RAYS

DETECTOR

a. Interrupted Light Sensor

REFLECTIVE SURFACE

EMITTER   DETECTOR

b. Reflective Light Sensor

INFRARED
LED

EMITTER

V

V_o
PHOTO-
TRANSISTOR

DETECTOR

c. Schematic

**Figure 3-13.
Photoelectric Position
Sensor**
*(Source: L. B. Masten and
B. R. Masten,* Understanding
Optronics, *Texas Instruments
Incorporated, Copyright ©
1981)*

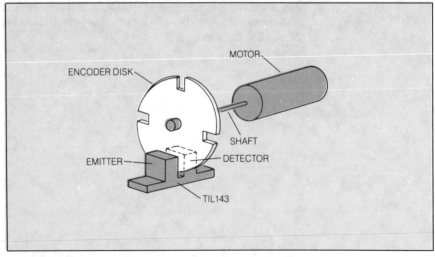

MOTOR

ENCODER DISK

SHAFT

EMITTER   DETECTOR

TIL143

Capacitive Switch

When detecting metal objects, the metal object can be used to form one plate of a capacitor and the change in capacitance generates an electrical signal.

One method of sensing a metallic object is to make it act as a plate of a capacitor. The sensor is shown in *Figure 3-14a*. As shown in *Figure 3-14b*, when no metal is present, the very small signal passed from the oscillator to the detector is due only to the very low leakage capacitance between plates of the device. When a metal object comes close to the sensor plates, two capacitors in series are formed as shown in *Figure 3-14c*. This capacitance is large by design and will pass a substantial portion of the oscillator signal. The sensitivity is dependent on the size of the sensor head and the distance between it and the metal object sensed.

Figure 3-14.
Metal Object Capacitive Sensor

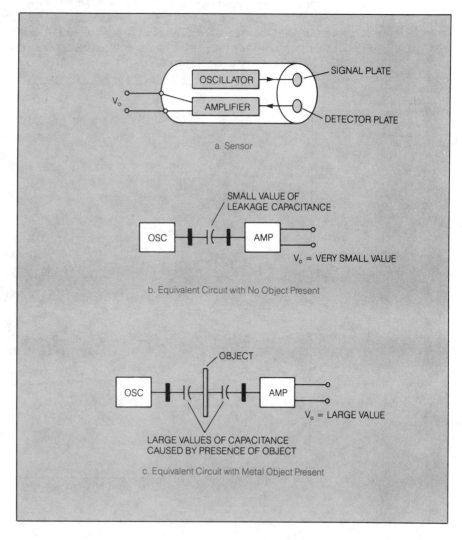

a. Sensor

b. Equivalent Circuit with No Object Present

c. Equivalent Circuit with Metal Object Present

Hall-Effect Switch

Hall-Effect switches are
made of semiconductor
material and generate a
voltage when current is
passing through the device
and the device is placed in
a magnetic field.

The Hall effect, which has been known for over a hundred years, is
the result of the force that is exerted on a charge carrier when in the presence
of a magnetic field. This is illustrated in *Figure 3-15*. A current I of charge
carriers (electrons) is passed through a small, thin, flat slab of semiconductor
material that has a magnetic field B passing through it. As a result, a voltage is
developed across the semiconductor material that is perpendicular to the
direction of current flow and perpendicular to the direction of the magnetic
field. By placing a magnet near a Hall-effect switch with current flowing
through it, the device will turn on (the voltage will be developed). This can be
used to sense position or a shaft rotation as shown in *Figure 3-16*.

One problem with the commonly available Hall-effect devices is that
amplification of the signal is required for use in an industrial environment,
especially if the signal must be transmitted more than a few feet. Shielding and
filtering also may be required.

Figure 3-15.
The Hall Effect
*(Source: W. B. Ribbens and
N. P. Mansour,* Understanding
Automotive Electronics, *Texas
Instruments Incorporated,
Copyright © 1982)*

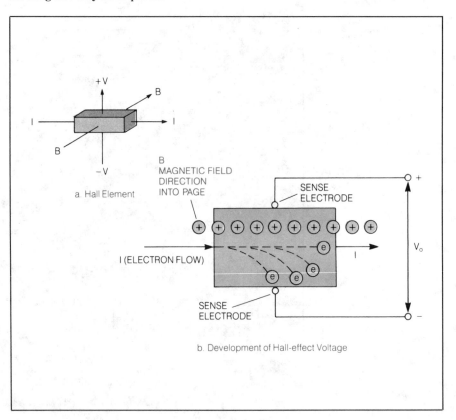

a. Hall Element

b. Development of Hall-effect Voltage

Figure 3-16.
Hall-Effect Position
Sensor
(Source: W. B. Ribbens and
N. P. Mansour, Understanding
Automotive Electronics, *Texas*
Instruments Incorporated,
Copyright © 1982)

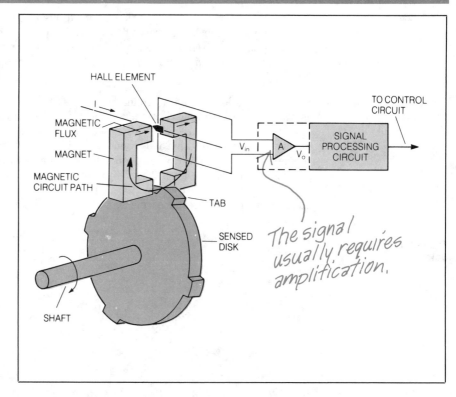

HALL ELEMENT

MAGNETIC FLUX

MAGNET

MAGNETIC CIRCUIT PATH

I

V_{in}

A

V_o

SIGNAL PROCESSING CIRCUIT

TO CONTROL CIRCUIT

TAB

SENSED DISK

The signal usually requires amplification.

SHAFT

Image Sensors

The most common image sensor method currently used digitizes the analog output of a TV camera so that a digital processor can use the information to position, aline or inspect parts.

Image sensing is an increasingly important parameter in industrial automation to replace the vision of the human eye. It is useful not only as a position finder, but also to perform inspection and alignment of parts. One method of accomplishing artificial vision is to digitize the analog output of a TV camera so it can be used by a digital processor. The digitized data is stored in memory. The data stored can be compared digitally to previously stored TV data to indicate a correct position, a correct object or some other action based on a visual decision. The individual details are fascinating but somewhat beyond the scope of this book. The importance and use of image sensors and image processing is definitely increasing in control systems.

ACTUATORS

Actuators provide the desired physical action by converting the electrical signal to a mechanical motion.

Both process control and discrete parts manufacture require a mechanical movement in order to accomplish a control function. Actuators are the devices that do the conversion from an electrical input to a mechanical action. In process control, this might be the setting of a valve to control the flow rate and/or amount of a chemical, heat, or gas input to a process. Discrete parts manufacturing requires the movement of the work piece or the work station in order to perform the assembly task. Actuators, therefore, are a vital part of any control system. Just as with sensors, research and development continue to produce better and more useful actuators.

Actuators may have a continuous or a discrete output. Continuous motion actuators usually are driven by an electric motor. Discrete motion actuators may be driven by special motors such as the stepping motor. Simple straight-line discrete motion can be provided by a solenoid.

Solenoid

Solenoids convert electrical energy (in the form of a magnetic field) to straight-line mechanical force for pulling or pushing.

The solenoid is an electromagnetic device that produces a straight-line mechanical force which is useful as an actuator. When S1 in *Figure 3-17a* is closed, the electrical current flowing through the solenoid coil produces a strong magnetic field within and around the coil. The magnetic field applies a force to the metal core to draw the core into the center of the coil. The core is normally held part way out of the coil by a spring (not shown). A device to be moved is attached to the core directly, or through mechanical levers to increase the force or the distance of movement. A simple application is shown in *Figure 3-17b*.

**Figure 3-17.
Solenoid**

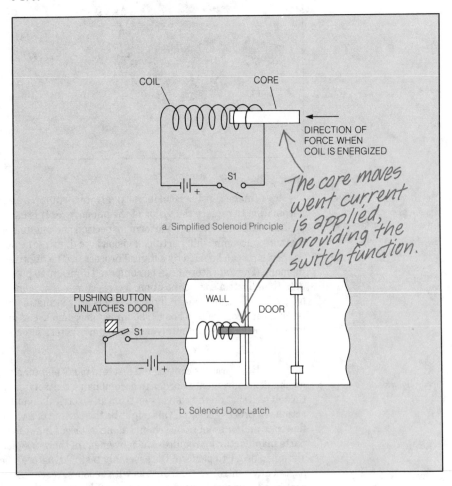

COIL   CORE

DIRECTION OF
FORCE WHEN
COIL IS ENERGIZED

S1

a. Simplified Solenoid Principle

*The core moves
went current
is applied,
providing the
switch function.*

PUSHING BUTTON
UNLATCHES DOOR   WALL   DOOR

S1

b. Solenoid Door Latch

Relay

The relay is similar in operation to the solenoid. The moveable armature is attracted to a fixed core and has a flexible finger with electrical contacts attached to it. The contacts open or close when the relay operates.

A relay is an important part of many control systems because it is an indirectly operated electrical switch that is useful for remote control and to control high current devices with a low current control signal. The same electromagnetic force principles of the solenoid are used, but the core is fixed in place. In *Figure 3-18*, the magnetic force pulls the armature toward the coil and, through the mechanical lever, forces the electrical contacts closed to close the controlled circuit. The springiness of the contact leaves opens the contacts when the coil is deenergized. Relays may have normally open contacts as shown, or normally closed contacts, or various combinations of both.

**Figure 3-18.
Simplified Relay**

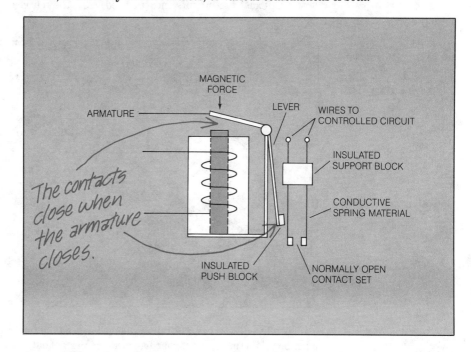

Electric Motors

Probably the most convenient of all actuators for control systems is the electric motor. The availability of low-cost solid-state electronics for control has resulted in many proportional actuators based on the electric motor. There are a number of different motor types, but we'll briefly discuss only four that are commonly used as actuators: the dc servo motor, the ac synchronous motor, the universal motor, and the stepping motor.

DC Servo Motor

The dc servo motor utilizes a variable dc voltage for speed control. A tachometer driven by the motor provides speed feedback to a control system called a servo loop and a rotational position sensor provides position feedback to form a self-contained positioning system. These motors are used where precise position control is necessary along with high-speed operation.

The basic dc motor structure is shown in *Figure 3-19*. It works on the principle that a force is applied to a current-carrying wire when the wire is located in a stationary magnetic field. The stationary magnetic field in the figure is provided by permanent magnets. The rotating armature winding (wires) is connected to the commutator segments and receives current through brushes rubbing against the commutator. The commutator and brushes form a switch for the armature current to keep the current flowing in the correct direction to provide continuous one-direction rotational torque. Much more information about electric motors is contained in another book in this Understanding Series™: D. L. Cannon, *Understanding Electronic Control of Energy*, Texas Instruments Incorporated, 1982, the source of *Figures 3-19* and *3-20*.

Figure 3-19.
2-Pole DC Motor Action
Source: D. L. Cannon,
Understanding Electronic
Control of Energy Systems,
Texas Instruments
Incorporated, Copyright ©
1982)

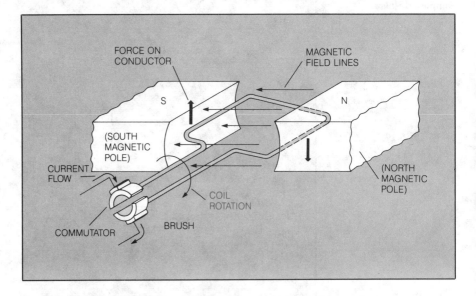

The dc servo motor can develop a large amount of torque, but care must be taken that it is not overdriven as the permanent magnets can be demagnetized. A substantial amount of control circuitry is required for a motor of this type, so it usually is not used unless its specific operating characteristics are required.

AC Synchronous Motor

The AC Synchronous
motor requires very little
control circuitry and is
less expensive and smaller
than an equivalent DC
motor, but is also limited
in its flexibility.

In an ac motor, current flows in the armature in the correct direction all the time and rotation is caused by a rotating magnetic field provided by an ac voltage. An ac synchronous motor requires very little control circuitry, but it is more limited in what it can do. It is useful in applications requiring continuous rotation, especially where a constant speed is required. Its speed is a function of the frequency of the input voltage and since it usually is connected to the utility power ac line which has a stable frequency, the motor speed is constant. *Figure 3-20* shows the basic structure of a three-phase ac synchronous motor.

**Figure 3-20.
Synchronous AC Motor
Action**
Source: D. L. Cannon,
Understanding Electronic
Control of Energy Systems,
Texas Instruments
Incorporated, Copyright ©
1982)

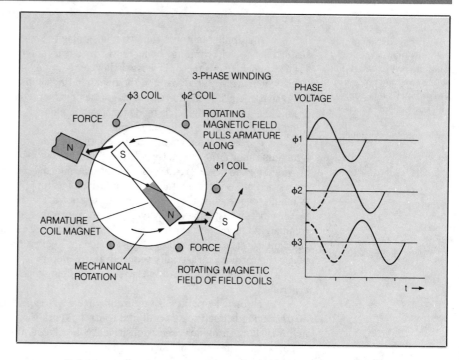

A three-phase synchronous motor is used where higher torque is required; typically, it is rated at more than one horsepower. It has advantages over a dc motor in that it is usually smaller, lighter, and less expensive than an equivalent horsepower dc motor. It can be used in proportional speed control applications if circuitry is provided to change frequency and relative phase. Single-phase and two-phase ac motors also are used in many applications.

The single-phase motor is not inherently self-starting and usually a small starting winding is powered through a capacitor phase-shifting network to develop a starting torque. The starting torque is poor, however, compared to other types of motors. To change direction of rotation, the phase of the starting winding voltage must be reversed which usually is accomplished by using external relays or switches. Because of this, this type motor normally is used in applications where reversing is not required.

Universal Motor

The universal motor will
operate on either AC or
DC voltage.

A universal motor gets its name from the fact that it will operate on either ac or dc voltage. This series-wound commutated-armature motor is very common and is found in widely varying applications. A series-wound motor is one in which the armature current flows through the field windings to develop the stationary magnetic field; therefore, permanent magnets are not required in this motor. It has a number of limitations, one of which is poor speed control, thus, it usually is used where speed is not critical. Sometimes it is used with a gear train or a clutch-brake mechanism.

Stepping Motor

When used with external circuitry to drive it, a stepping motor can be used with direct digital control. Due to its field windings, the stepping motor operates in a series of steps rather than a continuous motion.

A stepping motor is constructed similar to an ac synchronous motor. Instead of the field windings being driven by a continuously varying sine wave as an ac synchronous motor is driven, they are supplied with discrete voltage levels which are "rotated" around the field windings in increments or steps, thus the name. The result is that a magnetic field is incrementally rotated around the armature which is made up of permanent magnets. This causes the armature to rotate in a stepping fashion.

Figure 3-21 shows the basic structure of a dc stepping motor. The figure also shows the magnetic pole positions versus the coil digital signals provided by the drive circuitry. The rotational speed depends on how fast the magnetic field is incremented. One limitation of a stepping motor is that the available torque is inversely proportional to speed. Furthermore, a stepping motor must have its speed slowly increased from stop to full speed; if commanded to go to full speed from a full stop, it will stall immediately. In addition, most practical motors have a resonance point at a fairly low speed at which the torque is drastically reduced. If the motor is kept at this point very long, it will stall. Thus, the application of a stepping motor must be carefully controlled.

External circuitry is required to drive this type of motor, but the operation lends itself to direct digital control. Another advantage of a stepping motor is that it provides precise positional control of movement. This is an important requirement of machines that use indexed movement.

In these first three chapters, the concepts and fundamentals of control systems have been discussed. In future chapters, we'll see how these are applied to control systems in the real world.

**Figure 3-21.
Stepping Motor
Operation**

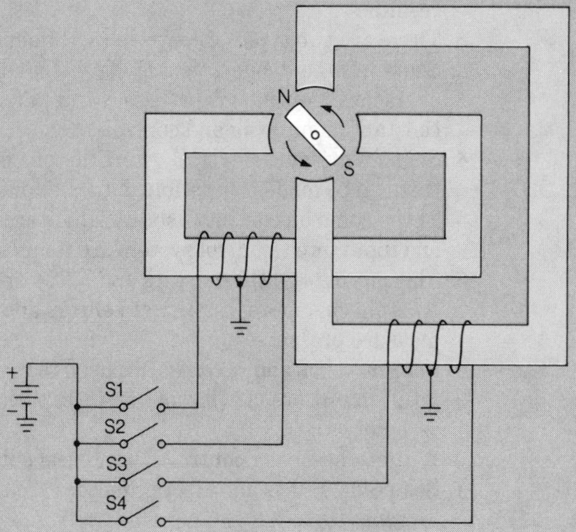

Code	S1	S3	S2	S4
1	0	0	1	1
2	1	0	0	1
3	1	1	0	0
4	0	1	1	0

1 indicates switch closed
0 indicates switch opened

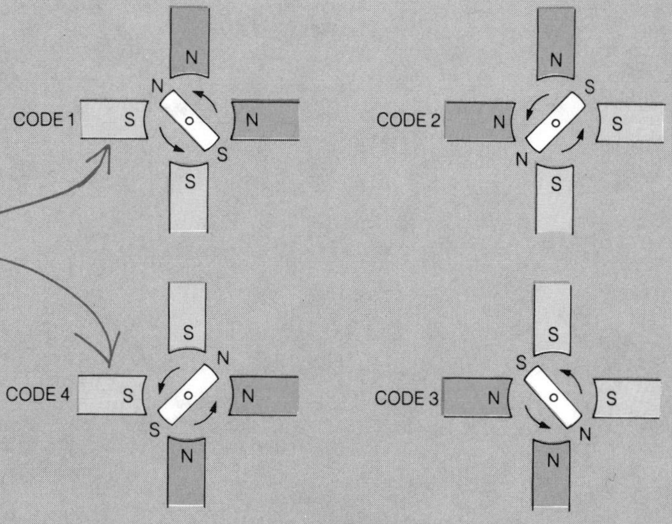

The magnetic field is rotated in increments.

WHAT HAVE WE LEARNED?

1. A controller receives information about a process, makes a decision based on this information, performs some action on the process, and monitors the results.
2. There are two types of controllers; sequential and continuous process.
3. Sequential controllers usually have ON/OFF signals and step-by-step operation. Continuous process controllers control a process that requires constant monitoring and control.
4. Sensors measure physical quantities of a process and provide an output that can be made compatible with the controller's input.
5. Three common continuously variable parameters that need to be measured and input into a control system are temperature, pressure and flow.
6. Thermocouples, thermistors and RTDs are used to measure temperature.
7. Aneroid chambers, capacitive sensors and strain gauges are used to measure pressure.
8. Venturi tubes and turbines are used to measure flow.
9. Position sensors use photoelectric, capacitance, Hall effect and imaging techniques.
10. Actuators convert controller electrical output signals into physical action.
11. Solenoids, relays and electric motors are the prime actuators in control systems.

Quiz for Chapter 3

1. Sensors are used to:
   a. measure temperature.
   b. measure flow.
   c. measure position.
   d. measure pressure.
   e. all of the above.

2. The output of a thermocouple is:
   a. a resistance.
   b. a capacitance.
   c. a current.
   d. a voltage.

3. A thermistor:
   a. is a pressure sensing device.
   b. has a voltage output.
   c. measures temperature.

4. Pressure sensors:
   a. use an aneroid chamber with a variable resistance output.
   b. use the piezoresistive effect in strain gauge sensors.
   c. use capacitive variations to sense pressure.
   d. all of the above.

5. A Hall effect switch is usually used to:
   a. measure temperature.
   b. measure voltage.
   c. sense a vacuum.
   d. sense the presence of a magnetic object.

6. Actuators are used to:
   a. activate a chemical.
   b. sense an object.
   c. make a mechanical movement.

7. A primary actuator component is:
   a. an ac synchronous motor.
   b. a universal motor.
   c. a solenoid.
   d. a stepping motor.
   e. all of the above.

8. A stepping motor:
   a. has a commutator.
   b. has a speed proportional to voltage input.
   c. is a good choice where continuous rotation is needed.
   d. can index positional movement.

9. An ac synchronous motor:
   a. can be reversed easily.
   b. can run at any speed by changing voltage.
   c. cannot stall.
   d. is useful when a constant speed motor is required.

10. A process controller:
    a. is useful in a continuous process control application.
    b. can interface to digital and analog I/O.
    c. can solve a PID process control equation easily.
    d. all of the above.

11. A drum timer is a good example of:
    a. an input sensor.
    b. an output actuator.
    c. a position sensor.
    d. a process controller.
    e. a sequence controller.

12. Light sensors:
    a. cannot be used as limit switches.
    b. come in a wide variety of mounting styles.
    c. work in direct sunlight.
    d. never use infrared light as light source.
    e. all of the above.

13. A sequence controller:
    a. is a mechanical only controller.
    b. is difficult to program.
    c. is no longer in widespread use.
    d. performs on/off control functions in relation to time.

14. Proportional control provides:
    a. discrete two-position control.
    b. control that varies with time.
    c. control that responds in discrete steps.
    d. continuously variable control.
    e. all of the above.

15. An aneroid barometer.
    a. can measure only atmospheric pressure.
    b. is very unreliable.
    c. replaces mercury thermometers in machine control.
    d. measures pressure by sensing deflection of an evacuated chamber.
    e. none of the above.

16. A venturi tube:
    a. is made only of metal.
    b. is extremely difficult to use.
    c. measures the presence of a liquid.
    d. can be used to measure gas or liquid flow rates.
    e. none of the above.

17. Stepping motors:
    a. can be built to index different amounts by changing the pole configuration.
    b. can run at varying speeds.
    c. can run both clockwise and counterclockwise.
    d. do not require commutators.
    e. all of the above.
    f. none of the above.

18. A dc servo motor:
    a. uses a variable dc voltage for speed control.
    b. uses a tachometer for speed feedback.
    c. can provide precise position control.
    d. all of the above.
    e. none of the above.

19. A solenoid:
    a. is spring operated.
    b. can be used only in dc circuits.
    c. is widely used in industrial automation.
    d. is related to an aneroid.
    e. none of the above.

20. A relay is:
    a. an indirectly operated electrical switch.
    b. widely used in industrial control automation.
    c. magnetically operated.
    d. all of the above.
    e. none of the above.

Basic Electronic Functions

ABOUT THIS CHAPTER

Originally, most industrial control systems were non-electronic in nature. They contained pneumatic, hydraulic, or purely mechanical controllers. Relays were used in later systems to assist in the control function and even today many systems are of this type. These systems are very durable, but are difficult to maintain and keep in calibration. Also it is very expensive and time consuming to make changes in the decision making process, if it is possible at all.

The rapid development of solid-state electronics from the transistor to the integrated circuit has changed and continues to change the approach and techniques used for automatic control systems. Digital logic, microprocessors, and microcomputers for decision making; analog amplifiers for amplification; and analog to digital converters and digital to analog converters for signal conditioning are a few of the basic elements in integrated circuit form. Let's look at how these developed.

THE TRANSISTOR

The transistor is one of the most important and fundamental inventions in the electronics industry. An extremely versatile device, it is used in digital logic and memory and analog amplifier applications.

The foundation of modern solid-state electronics is the transistor. It was invented in 1947 at Bell Laboratories and permanently changed the electronics industry. The transistor is used as a switch in digital logic circuits to make decisions; used in memory devices to store information; and used in analog circuits to amplify and shape an electrical signal. A transistor is a semiconductor because it is made from material such as silicon that is midway between a good conductor of electricity like copper and a good insulator like rubber.

Figure 4-1a shows a transistor that can be used as a switch. The silicon is of two types, an n-type and a p-type, which are made by adding different impurities to the silicon and which react differently electrically. A basic law of semiconductors states that electricity cannot flow across a junction from an n-type to a p-type material if the n-type has a voltage applied to it that is more positive than the p-type. The junction is said to be "reverse biased".

Figure 4-1.
A Transistor Model
(Source: Gene McWhorter,
Understanding Digital
Electronics, *Texas Instruments*
Incorporated, Copyright ©
1978)

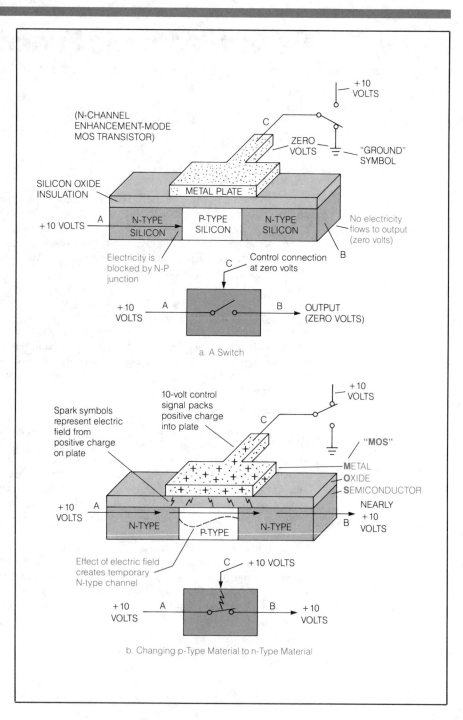

a. A Switch

b. Changing p-Type Material to n-Type Material

An MOS transistor is made of a sandwich of semiconductor material with either N or P type diffusions, a silicon oxide insulator, and a metal plate used to control current in the transistor.

Therefore, with +10 volts applied to terminal A, no current will flow to terminal B because it is blocked by the n-p junction. The transistor is said to be in an "off" or "open" state. However, if +10 volts is applied to the metal plate C that is above the silicon and separated from it by an insulator (silicon oxide), the p-type material is effectively changed to an n-type material by an electric field from the metal plate as shown in *Figure 4-1b*. Now current can flow between terminal A and terminal B and, as shown in *Figure 4-1b*, terminal B has nearly +10 volts on it. In this condition, the transistor is said to be in an "on" or "closed" state. Thus, the transistor can be made to act as a switch by controlling the voltage on the metal plate. Such a device is known as an MOS transistor for the sandwich of material that it's made from—Metal, Oxide, Semiconductor.

THE INTEGRATED CIRCUIT

By combining large numbers of transistors, diodes and resistors on one tiny piece of silicon, the integrated circuit (IC) has made possible the development of microprocessors and microcomputer controllers.

The next major breakthrough occurred at Texas Instruments in 1958 where a new method was developed that allowed multiple numbers of interconnected transistors, diodes, and resistors to be made all at the same time in one piece of silicon. This new device was called an integrated circuit (IC). Today the piece of silicon out of which the circuit is made often is called simply a "chip". Technology has advanced to the point where circuits with over 100,000 transistors, along with other devices such as resistors and capacitors, can be placed on a chip approximately ¼ inch (6.35 millimeters) square. Many hundreds of identical circuits are made at the same time on a slice of silicon having a diameter of between three and six inches (7.6 and 15.2 centimeters). These components are interconnected to perform some function by a pattern of metal conductors evaporated on the chip surface and then selectively etched. *Figure 4-2* shows an IC interconnected to operate as a microprocessor.

All the devices to be discussed in this chapter are available in integrated circuits. Many are simple devices that have far fewer than 100,000 transistors, but are manufactured by this important technique which has reduced the size, power consumption, and cost while increasing the performance, available features and reliability of electronic products.

COMBINATIONAL LOGIC GATES

Combinational logic gates have an output that is based on the specific condition of present inputs.

Combinational logic gates are used to make decisions in digital systems based on present events. They do not remember what decisions they have previously made. A logic gate is designed to have a specific output only if certain conditions are present at the input to the logic gate. The inputs and outputs of logic gates are defined with two levels, the binary 1 and 0 that were discussed previously. For the purpose of the following discussion, a binary 1 will be considered an "on" condition and a binary 0 will be considered an "off" condition. Let's discuss the three basic logic functions upon which all combinational logic gates are built: the AND, OR, and NOT functions.

Figure 4-2.
A Typical Integrated
Circuit
Source: D. L. Cannon and
G. Luecke, Understanding
Microprocessors, *Texas*
Instruments Incorporated,
Copyright © 1979)

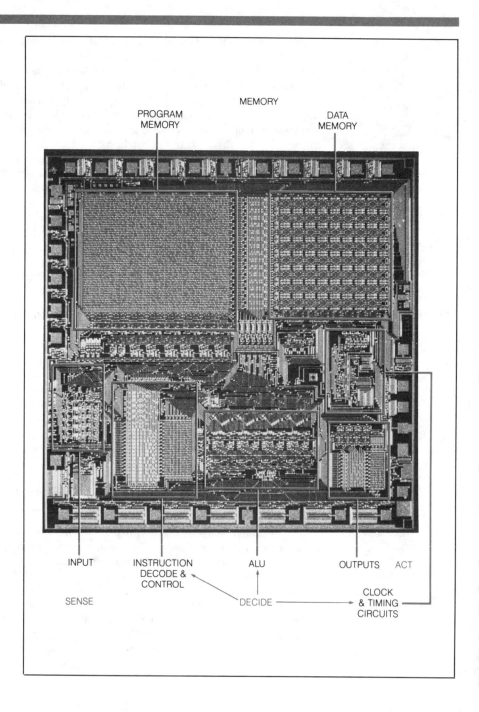

The AND Gate

AND gates are logic gates in which all inputs must be a logic 1 for the output to be a 1, otherwise, the output is always 0. They can be used in control processes to indicate that all steps of a specific operation have been completed.

Figure 4-3a shows a circuit in which the lamp will be lit only if switch A AND B are closed. This is a functional representation of the AND gate. A table relating whether the lamp is "on" or "off" in relation to the switch positions is shown in *Figure 4-3b*. The AND gate symbol used on logic diagrams is shown in *Figure 4-3c*. In all logic diagrams using symbols like *Figure 4-3c*, it is assumed that supply voltage(s) and ground are connected to the circuit. The table in *Figure 4-3d* is called a truth table and indicates the output for all the possible combinations of the inputs; in this case, all inputs must be 1 for the output to be 1. The AND gate may have more than two inputs, but the same law applies:

If all inputs of an AND gate are a logic 1, then the output is a logic 1; otherwise, the output is a logic 0.

Recall that +5 volts often is used to indicate a binary 1 and zero volts to indicate a binary 0; so the output will be +5 volts if, and only if, both inputs are +5 volts. There is nothing magic about +5 and 0 volts; other values of voltages could be chosen. For example, +9 volts could be a binary 1 and +1.88 volts could be a binary 0. However, to make circuits and systems work together, it must be agreed in advance that particular values will be used and +5 and 0 volts are commonly used and accepted levels.

Figure 4-3.
The AND Operation,
Symbol and Truth Table

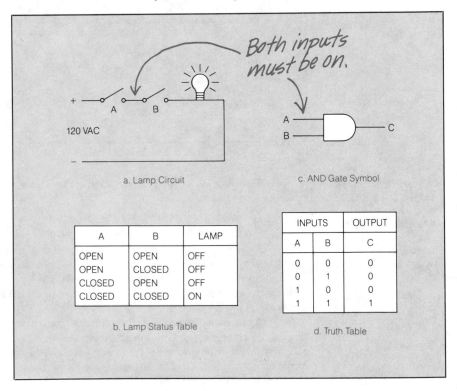

a. Lamp Circuit

c. AND Gate Symbol

A	B	LAMP
OPEN	OPEN	OFF
OPEN	CLOSED	OFF
CLOSED	OPEN	OFF
CLOSED	CLOSED	ON

b. Lamp Status Table

INPUTS		OUTPUT
A	B	C
0	0	0
0	1	0
1	0	0
1	1	1

d. Truth Table

A practical use of an AND gate in process control is shown in *Figure 4-4*. The control system is designed so that a 1 is present at each input when the condition is true. If none or only one or two conditions are true, the lamp is not lit. If all three are true, the lamp is lit. Thus, the lamp indicates the status of this point in the process. This may be for information only or for instructing the operator to proceed with the next step. The output also could be connected to another circuit which would automatically allow the process to proceed to the next step when all input conditions are as specified.

**Figure 4-4.
Using an AND Gate in an
Industrial Process**

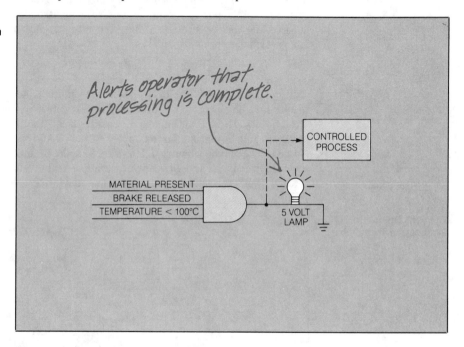

The OR Gate

When any input is a logic 1, the OR gate's output will be a logic 1. Since all inputs must be a logic 0 before the output is logic 0, the OR gate is well suited to monitor alarm systems.

The basic law governing operation of the OR logic gate can be stated as:

If any one (or more) input to an OR gate is a logic 1, then the output is a logic 1; otherwise, the output is a logic 0.

Figure 4-5a shows a circuit that is a functional representation of an OR gate. If either switch A OR switch B is closed or both are closed, then the light is lit as indicated in the table of *Figure 4-5b*. The logic symbol and truth table for the OR gate are given in *Figures 4-5c* and *4-5d*. Again, more than two inputs are allowed.

A common use for OR gates in process control is to monitor conditions that call for emergency action. For instance, as shown in *Figure 4-5e*, the OR gate may monitor for smoke, loss of power, excessive temperature or an intruder in the room. All of these signals are connected to one OR gate. The presence of one or more of these conditions causes the OR gate to generate an output which sounds an alarm.

Figure 4-5.
The OR Operation,
Symbol and Truth Table

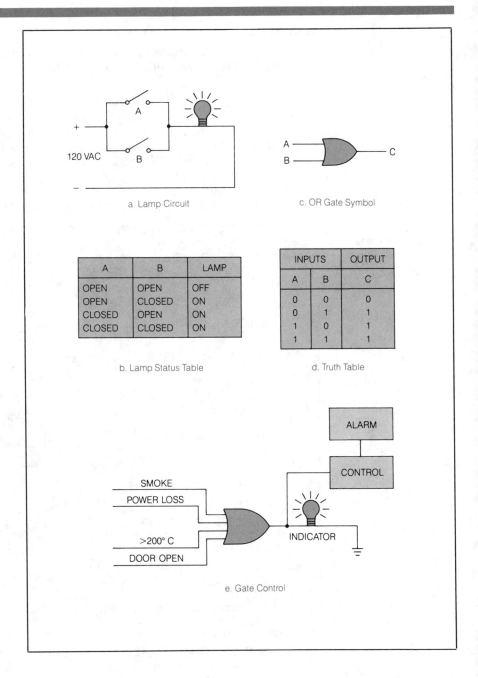

a. Lamp Circuit

c. OR Gate Symbol

A	B	LAMP
OPEN	OPEN	OFF
OPEN	CLOSED	ON
CLOSED	OPEN	ON
CLOSED	CLOSED	ON

b. Lamp Status Table

INPUTS		OUTPUT
A	B	C
0	0	0
0	1	1
1	0	1
1	1	1

d. Truth Table

e. Gate Control

The NOT Gate

The output of a NOT gate inverter is always the opposite of the input. It inverts the signal.

The last of the three basic logic functions is called the NOT function and is described in *Figure 4-6*. The law for its operation is:

When the input to a NOT gate is a logic 1, the output is a logic 0; when the input is a logic 0, the output is a logic 1.

The NOT gate also is called an inverter because the output is always the opposite of the input. Unlike other gates, this gate can have only one input.

Figure 4-6a shows the electric circuit representation of the NOT function. If switch A is open, the lamp is on; if switch A is closed the lamp is off as indicated in *Figure 4-6b*. The resistor, R, is included to prevent a short circuit across the 120 volt input when switch A is closed. *Figure 4-6c* is the electronic symbol for the NOT gate and *Figure 4-6d* shows its truth table.

**Figure 4-6.
The NOT Operation,
Symbol and Truth Table**

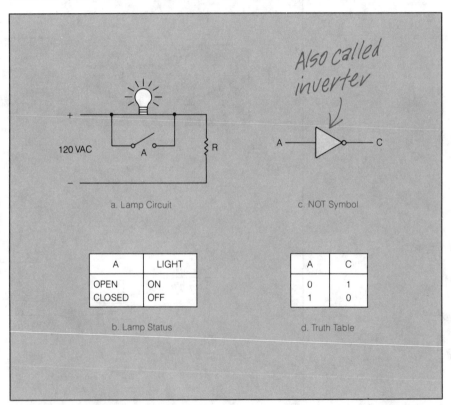

Also called inverter

a. Lamp Circuit

c. NOT Symbol

A	LIGHT
OPEN	ON
CLOSED	OFF

b. Lamp Status

A	C
0	1
1	0

d. Truth Table

The NAND and NOR Gates

The three basic logic functions can be combined to form two more logic functions that are packaged as one complete circuit. *Figure 4-7* shows a combination of the AND and NOT gates to form a NAND gate. Compare the truth tables of *Figure 4-3d* and *Figure 4-7d* to verify that the output has been negated in *Figure 4-7d* as a result of the NOT gate acting on the output of the AND gate.

**Figure 4-7.
The NAND Gate**

By combining the basic
NOT, AND and OR logic
functions, two important,
additional logic gates can
be constructed: The
NAND (Not and AND
gates combined); and the
NOR (OR and NOT gates
combined).

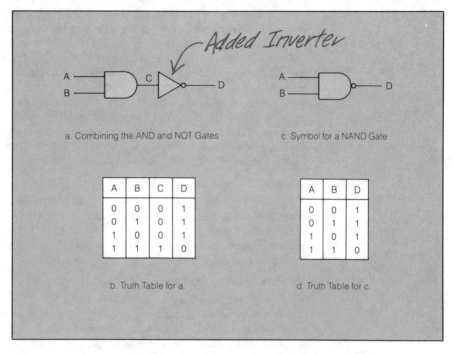

a. Combining the AND and NOT Gates

c. Symbol for a NAND Gate

A	B	C	D
0	0	0	1
0	1	0	1
1	0	0	1
1	1	1	0

b. Truth Table for a.

A	B	D
0	0	1
0	1	1
1	0	1
1	1	0

d. Truth Table for c.

Figure 4-8 shows that a NOR gate is formed by combining an OR gate
and a NOT gate. Compare the truth tables of *Figure 4-5d* and *Figure 4-8d*.

**Figure 4-8.
The NOR Gate**

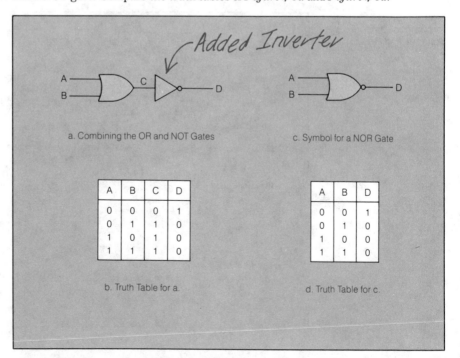

a. Combining the OR and NOT Gates

c. Symbol for a NOR Gate

A	B	C	D
0	0	0	1
0	1	1	0
1	0	1	0
1	1	1	0

b. Truth Table for a.

A	B	D
0	0	1
0	1	0
1	0	0
1	1	0

d. Truth Table for c.

Five logical functions have been defined that are commonly available as complete circuits to be used in digital logic circuits to make decisions. It must be emphasized again that these circuits do not remember what decisions they make — they do not have any memory capability. Their output at a particular time is based only on the condition of the inputs. Study the operation of the gates carefully and understand how they work because these gates will be used shortly in electronic building blocks.

The methods for designing systems using these gates is beyond the scope of this book. Further information of this type can be obtained by reading another book in this Understanding Series™; Gene McWhorter, *Understanding Digital Electronics*, Texas Instruments Inc., 1978.

COMBINATIONAL ELECTRONIC BUILDING BLOCKS

Putting combinational logic gates together in various ways on an integrated circuit results in what are called electronic building blocks. These, along with the basic gates, can be used to perform tasks in an automated system. Three popular devices will be discussed: the data selector/router, the encoder, and the decoder.

The Data Selector/Router

Often in electronic control applications, there is a requirement to select only one of several input signals and pass it on. Likewise, there may be a need to route one signal to any one of several outputs. *Figure 4-9* shows these conditions in schematic form using a manually controlled 4-position switch. For electronic control of the switching, the electronic data selector and data router are used.

**Figure 4-9.
Mechanical Data
Selector and Router**

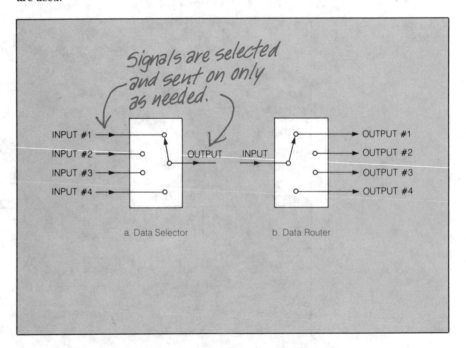

Signals are selected and sent on only as needed.

INPUT #1

INPUT #2   OUTPUT   INPUT

INPUT #3

INPUT #4

a. Data Selector

OUTPUT #1

OUTPUT #2

OUTPUT #3

OUTPUT #4

b. Data Router

Data Selector/Router integrated circuits are constructed using combinational logic gates that use a control code to select 1 of n inputs to send to an output or selects 1 of n outputs to receive an input signal.

Figure 4-10.
Symbols and Code Table for the Data Selector and Data Router

The symbols for the electronic data selector and data router are shown in *Figure 4-10a* and *Figure 4-10b*, respectively. Coded input signals to the two control lines determine which input (or output) is to be selected. The number of control lines needed depend on the number of inputs (or outputs) because n control lines can select any one of 2^n inputs (or outputs). Thus, two control lines are needed for four inputs (or outputs). The code table in *Figure 4-10c* gives one possible coding method to select one of the four I/O lines. From the table, if C1 = 1 and C2 = 0, then I/O line 3 is connected.

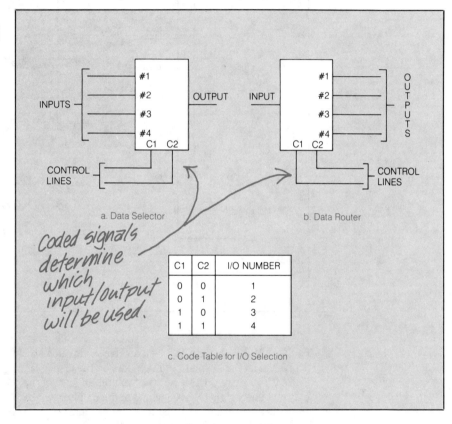

a. Data Selector

b. Data Router

Coded signals determine which input/output will be used.

C1	C2	I/O NUMBER
0	0	1
0	1	2
1	0	3
1	1	4

c. Code Table for I/O Selection

The example data selector can be built using basic logic gates as shown in *Figure 4-11*. Consider input number 2. For it to be connected to the output, the other two inputs to AND gate B must be logic 1. One of these inputs is C2 direct and the other is the inversion of C1 because of the NOT gate. Therefore, if C1 = 0 and C2 = 1, the two control inputs into gate B are both 1. Then, if the input number 2 signal is 0, the output is 0. If the input number 2 signal is 1, the output is 1. That is, input number 2 is essentially "connected" to the output since the output follows input number 2. (Actually, it isn't connected in the way a mechanical switch would be, but the effect is the same.) Take the time to understand that only one input can be connected to the output OR gate at any given time and that the output will be at the same level as the gated input.

Figure 4-11.
Basic Logic Gates
Combined to Make a
Data Selector

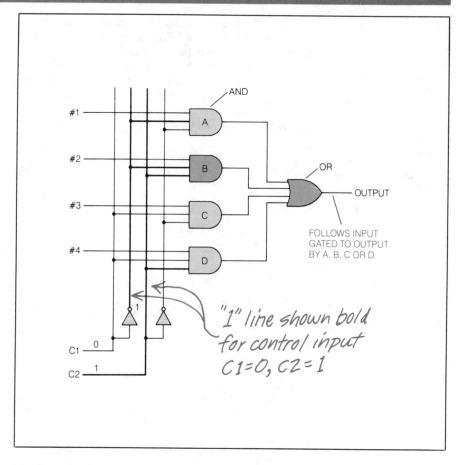

The Encoder/Decoder

The encoder changes incoming signals into a binary code that can be used by the computer. The decoder changes the computer binary code back into a form required by a receiving device.

Most everyone has used an electronic calculator. When the "5" button is pressed on the keypad, a "5" appears on the display and is entered into the calculating circuits. However, the "computer" part of the calculator operates only on binary signals, not decimal numbers like the "5". Therefore, the "5" from the keyboard must be converted to a binary signal, called a binary code, that represents the decimal number. This is the job of an encoder. It must realize when a key has been depressed, determine which key it is, and generate the binary code for that key before a second key is depressed. The function of a decoder is just the opposite — to receive a binary number code from the computer and decode that number into the form needed to drive the display unit to indicate the decimal number. These steps are shown in *Figure 4-12.*

**Figure 4-12.
Encoder and Decoder
Used in Computer
System**

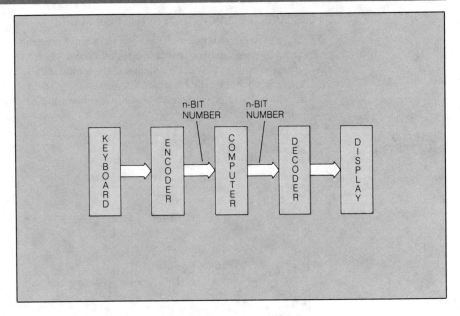

The encoder generates a unique n-bit code for each key. For instance, if $n = 3$, the encoder can accommodate eight keys and a 3-bit code will be generated for each key. The assignment code may be any one of several, but a possible one is given in *Table 4-1*.

**Table 4-1.
Keypad Code**

Key No.	Encoder Output
1	000
2	001
3	010
4	011
5	100
6	101
7	110
8	111

A binary digital code the computer understands.

The encoder in control systems may accept a variety of input codes, and the decoder may drive a variety of output devices.

Thus, if key number 4 is depressed, the encoder output will be 011. The computer then uses 011 to represent decimal number 4 in its calculations.

In similar fashion, the computer sends an n-bit binary code to the decoder which converts it into an output compatible with the display being used. Different displays require different inputs and the decoder must be selected to match the display device. For an n-bit input, the decoder must generate 2^n unique outputs.

In process control, the "keyboard" may consist of a simple 4-button keypad or a full typewriter-type keyboard. The display units could be lights, dot-matrix displays, 7-segment displays, audio tones, printer, CRT, etc.

SEQUENTIAL LOGIC GATES: THE FLIP-FLOP

In many applications, the system must remember what decisions it has made previously so that the output will be a function of these previous actions as well as the current values of the input signals. Logic devices with this capability are called "sequential" or "memory" devices. The basic sequential logic gate is the simple flip-flop shown in *Figure 4-13* along with its truth table.

**Figure 4-13.
Basic Flip-Flop**

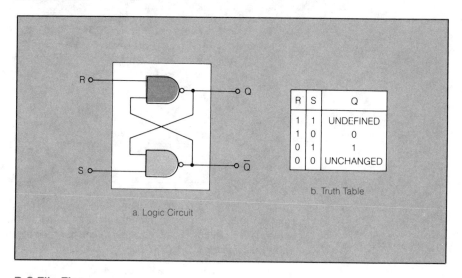

R	S	Q
1	1	UNDEFINED
1	0	0
0	1	1
0	0	UNCHANGED

b. Truth Table

a. Logic Circuit

R-S Flip-Flop

The R-S Flip-Flop (Reset-Set Flip-Flop) is formed by combining two NAND gates that are cross-coupled. In its simplest form, it has an output state that can be indeterminant if both R and S are 1's at the same time.

The R-S flip-flop consists of two cross-coupled NAND gates. R and S stand for reset and set, respectively. If S goes to a 1 while R is 0, the output, Q, is "set" to 1. If R goes to a 1 while S is 0, the output, Q, is "reset" to 0. In either case, when R (or S) subsequently returns to a 0 after being a 1 (and the other input stays at a 0), the flip-flop remembers that R (or S) previously was a 1 and the output, Q, does not change. Thus, only when R or S goes to a 1 can the output, Q, change. The output $\overline{Q}$ (read as "Q not") is the NOT or inverse of the output Q.

There are two problems with the R-S flip-flop. If R and S both become 1 at the same time, the output is uncertain. This is called an undefined or unstable state. Another problem is that the flip-flop operates asynchronously; that is, the output changes *whenever* R or S goes to 1. This is a problem when the output should change only at predetermined times. This changing at specific times, called synchronous operation, is accomplished by adding a C (clock) input and two more NAND gates as shown in *Figure 4-14*. Also shown are two more control lines called "preset" and "clear". These allow separate control to preset the output to a 1 or 0 at any time. The truth table and symbol for the device also are given. In synchronous operation, the flip-flop responds to the inputs only when the clock signal is present. Inputs are indicated at time period n when the clock arrives; outputs are indicated at time period n + 1 when the circuit has responded.

Figure 4-14.
R-S Synchronous Flip-
Flop With Preset/Clear

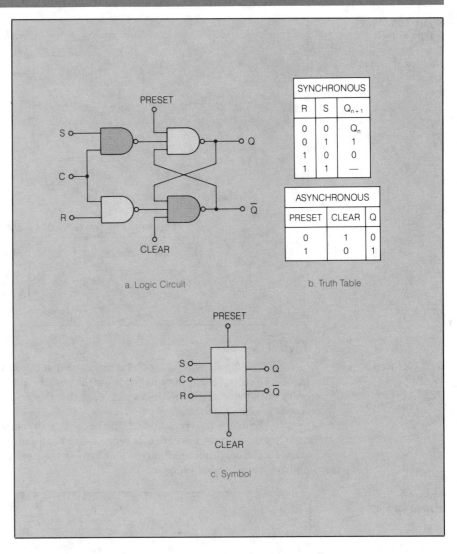

a. Logic Circuit

b. Truth Table

SYNCHRONOUS		
R	S	Q_{n+1}
0	0	Q_n
0	1	1
1	0	0
1	1	—

ASYNCHRONOUS		
PRESET	CLEAR	Q
0	1	0
1	0	1

c. Symbol

D-Type Flip-Flop

The D-type of flip-flop is also called a "latch," it catches the data at the input at clock time and holds it or "latches" it for a relatively long time.

If the R-S flip-flop is modified as indicated in *Figure 4-15*, the result is called a D-type flip-flop, or more commonly a *latch*. Notice that the reset input is generated by passing the set input through a NOT gate. The truth table also is shown and indicates that the output Q assumes the value of the set input every time a clock pulse is received. The set input can change, but Q remains latched and will not change until the next clock pulse is received.

Latches are used to capture binary information that may be present only for a few microseconds, but where the device that will use the information needs the data available for a much longer period of time.

Figure 4-15.
D-Type Flip-Flop

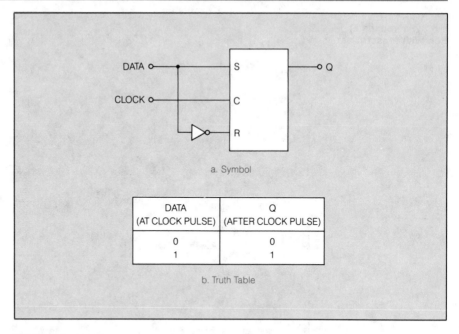

a. Symbol

DATA (AT CLOCK PULSE)	Q (AFTER CLOCK PULSE)
0	0
1	1

b. Truth Table

T-Type Flip-Flop

The T-type flip-flop "toggles" when it receives a clock pulse i.e. its output flips to the opposite state.

The flip-flop in *Figure 4-16* is unique in that the only input is the clock pulse. The output changes value (state) on the positive-going or leading edge of every clock pulse. Digital systems people say that the flip-flop "toggles" upon receipt of every clock pulse.

Examine the truth table. C_1 is the value of the clock input at time t_1 and C_2 is the value at t_2 where t_2 is very close to t_1. Q_1 and Q_2 are the outputs at t_1 and t_2. The only time Q_2 changes value is when the clock pulse has had a positive transition between t_1 and t_2.

Figure 4-16.
T-Type Flip-Flop

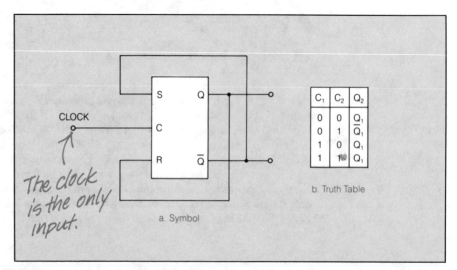

The clock is the only input.

a. Symbol

C_1	C_2	Q_2
0	0	Q_1
0	1	$\overline{Q_1}$
1	0	Q_1
1	1	Q_1

b. Truth Table

Figure 4-17 shows a clock input and the resultant T-type flip-flop output. Notice that the output changes half as often as the input; thus, the frequency of the clock signal has effectively been divided by two. If this flip-flop output were fed to the clock input of another T-type flip-flop, the output of the second flip-flop would be only one-fourth the original clock frequency. Thus, this device is useful in divider circuits and in counter circuits.

**Figure 4-17.
Output Q of a T-Type
Flip-Flop for Input C**

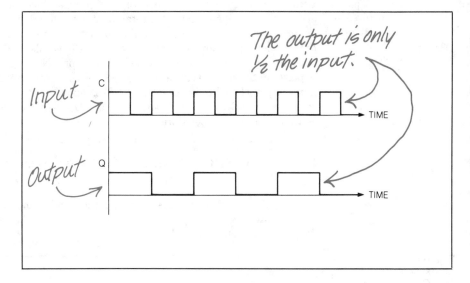

J-K Flip-Flop

The J-K flip-flop, unlike an R-S flip-flop, changes its output when both its inputs are a 1, avoiding the uncertainty this condition causes.

In order to overcome the uncertain condition when both R and S are 1 in the R-S flip-flop, the J-K flip-flop was developed. The symbol for this flip-flop and its associated truth table are shown in *Figure 4-18*. When both J and K are 1, the output changes state; otherwise, the operation is exactly like the basic R-S flip-flop.

Four basic types of flip-flops have been discussed: R-S, D, T, and J-K. There are many variations and combinations of these four types. The basic types and some of their variations will be used in the circuits that follow.

**Figure 4-18.
J-K Flip-Flop**

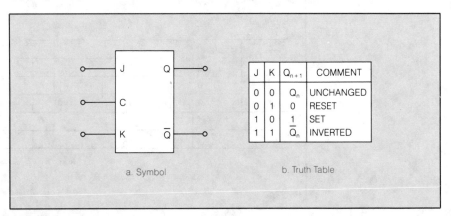

J	K	Q_{n+1}	COMMENT
0	0	Q_n	UNCHANGED
0	1	0	RESET
1	0	1	SET
1	1	$\overline{Q}_n$	INVERTED

a. Symbol   b. Truth Table

SEQUENTIAL ELECTRONIC BUILDING BLOCKS

One important sequential logic building block, the latch, has been discussed. Other basic building blocks are the counter, parallel register, and memory.

The Binary Counter

The total number of pulses that have been received by a device are counted by a network of flip-flops called a binary counter. The output is a binary code.

A binary counter is a network of flip-flops which counts pulses as they arrive at the input and stores the total count in the flip-flops. Outputs usually are provided so that the stored total can be determined. *Figure 4-19* shows a typical binary counter made of four T-type flip-flops. Also shown is the input pulse train and the resulting parallel binary output. Notice that when a total of 16 pulses (0 through 15) has been received, the count starts over at 0. The letters A, B, C, and D represent the 2^0, 2^1, 2^2 and 2^3 positions of the binary code, respectively. If A = 1, B = 1, C = 0, and D = 0 then the decimal equivalent of the number stored is:

$$0 \times 2^3 + 0 \times 2^2 + 1 \times 2^1 + 1 \times 2^0 = 2 + 1 = 3$$

Likewise, if A = 0, B = 1, C = 1, and D = 1, then the equivalent number is:

$$1 \times 2^3 + 1 \times 2^2 + 1 \times 2^1 + 0 \times 2^0 = 14$$

If only the D output is used, then the circuit shown is called a divide-by-16 counter because the output signal has a frequency that is 1/16 the frequency of the input signal.

Figure 4-19.
An Asynchronous Binary 4-bit Counter Made of "T" Flip-Flops
(Source: Gene McWhorter, Understanding Digital Electronics, Texas Instruments Incorporated, Copyright © 1978)

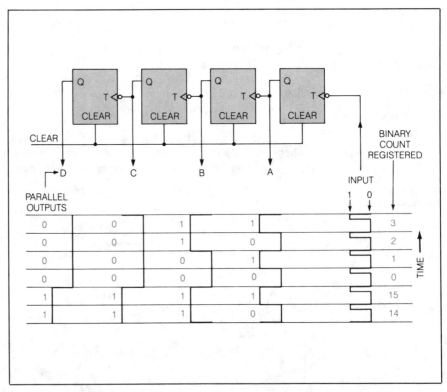

Parallel Registers

The parallel register provides a means for storing several bits of a digital code simultaneously. The stored data can be "read" without being destroyed.

Recall that the latch circuit can store one bit, but quite often a means for storing two or more bits of an n-bit binary word is needed. This can be done in a device called a parallel register, usually simply called a register. If 4 bits of data are to be stored simultaneously, four D-type flip-flops (latches) can be used, all of them clocked at the same time. They capture the data present at each of their inputs and store this data until instructed to capture new data. The stored data can be "read" by examining the Q output of each latch without destroying the data stored in the latch. This type of register is fundamental to the operation of computers and is used mainly in the processor itself for temporary storage of instructions and data as each instruction is executed. For longer storage time of large amounts of data, the memory circuits are used. If the four flip-flops in *Figure 4-19* were D-type flip-flops with an input for each bit and the clear line is changed to a clock line, then the parallel output of A, B, C and D would represent the outputs of a parallel register.

Memory Circuits

Electronic memory circuits are one of two basic types: Read-Only Memory (ROM) or Random Access Memory (RAM). ROM is non-volatile memory because the contents are not lost if power is disconnected from the ROM. Conversely, RAM is volatile memory because it loses the information stored in it when power is removed from the RAM.

Read-Only Memory

Instructions in binary code are built into a ROM (Read Only Memory) integrated circuit during its manufacture, and are designed to keep their program permanently even though power fails.

The basic kind of ROM is a mask programmable ROM that is manufactured with the binary information built into it. Because of the effort to design the tooling (masks), the first device costs several thousand dollars to build, but additional devices are relatively inexpensive. This method is practical only if several thousand of the same ROM are needed (such as in automotive or video game applications).

The second kind is programmed after it is manufactured. The programmable ROM or PROM permits the ROM to be programmed with inexpensive equipment one time after manufacture. This allows one-of-a-kind ROMs to be made cheaply. If a mistake is made in programming, however, the PROM must be discarded and the programming must be done on a new unit.

Since programming errors are common, another kind, an erasable PROM or EPROM was developed. The EPROM can be erased and reprogrammed in a short period of time using an ultraviolet light. The disadvantage is that the entire PROM must be erased.

The electronically alterable ROM or EAROM allows only selected portions of the ROM to be changed electronically. Because the EAROM is relatively expensive, the EPROM is the most popular type of ROM for use in developing controls for present day automation systems.

The most common usage for ROM is to store information that is likely to be used over and over without change and which needs to be preserved with power off. A program for a computer is a good example. For industrial controllers, the EPROM is usually used because the program is changed occasionally to control a modified or new operation. The EPROM is erased and reprogrammed to include the change.

As shown in *Figure 4-20*, the input lines to a ROM are chip select and multiple address lines. The chip select line allows selection between two or more ROM chips in the system. If the ROM contains 2^m memory locations, then m address lines are required to identify the locations to be read. Each memory location will store n bits of information, therefore, n data lines are required to simultaneously read all the bits of a location.

**Figure 4-20.
ROM Inputs and Outputs**

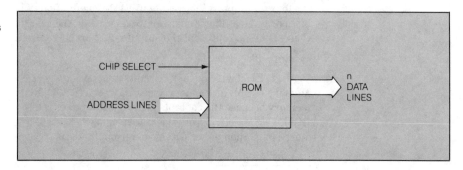

One popular ROM chip can store 16,384 (commonly called 16K) bits of information on the chip. One configuration may be to have 2,048 (2K) locations with 8 bits (a byte) stored at each location. In this case, 11 address lines (2^{11} = 2,048) and 8 data lines would be required. If four such chips are used in the system, 2 chip select lines would be needed to choose any one of the four chips. Four chips would provide storage for 8,192 8-bit words (8K bytes).

Random Access Memory

RAM (Random Access Memory) integrated circuits are similar to ROMS except that data can be stored (written) into and retrieved (read) from any location as desired.

Often data must be stored temporarily for later retrieval and use, then after use, the data is changed. The data may originate from sensors or may be intermediate results of calculations. The data also could be results of an analysis that will be sent to an I/O device such as a display or printer. This type of data is stored in a RAM.

A RAM is more complicated than a ROM since it must be capable of storing (writing) data that changes quickly and often, as well as reading the data that has been stored. However, in terms of inputs and outputs, it is very similar to the ROM. One input line must be added that tells the RAM whether data is to be written or read. This line is appropriately called the read/write (R/W) line. When the R/W line is at one level, the data on the data lines is to be stored (written) into memory; if at the other level, the data on the data lines is information being read from memory. Each memory bit location is a D-type flip-flop so that data present on the data lines can be latched in for a write operation. The state of the R/W line determines whether the latch inputs or outputs have access to the data lines.

Three RAM chips are shown connected to data lines called a data bus in *Figure 4-21*. Although all three have physical wiring connections to the data bus, only one chip can be electronically connected to the data bus at any given time. As a result, only one chip is allowed to output data from the address on the address bus at any one time.

Figure 4-21.
Three-Chip RAM System

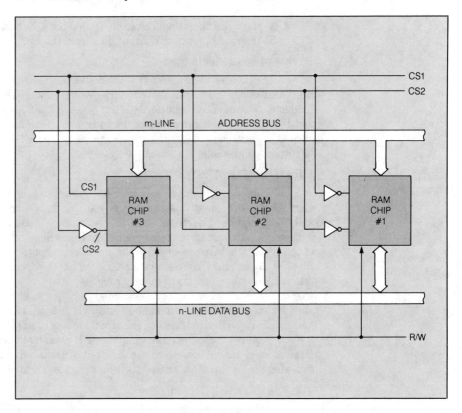

The chip select lines control which chip is attached to the buses. If it is assumed that a chip is selected when the CS inputs to the RAM chip are 1, then for *Figure 4-21*, the following assignment has been made:

Chip Select Line		RAM Chip Attached to Buses
#1	#2	
0	0	#1
0	1	#2
1	0	#3

Therefore, to use a RAM, follow these steps:

1. Select a chip by sending the proper binary code on the chip select lines.
2. Send the address of the memory location by placing its binary code on the address bus.
3. Send a signal on the read/write line to indicate which operation to perform.
4. a. If a write operation, place the data on the data bus.
   b. If a read operation, read the data from the data bus.

Review of Sequential Logic

Sequential logic means that the device output depends on the previous value of the output as well as the present inputs. The flip-flop is the fundamental sequential logic device. Four types of flip-flops are the R-S, D, T, and J-K. These basic devices can be used to make electronic building blocks such as latches, counters, registers, and memory. These building blocks are important components of automation systems.

LINEAR INTEGRATED CIRCUITS

Integrated Circuits which use analog instead of digital signals are called linear circuits. The operational amplifier (op amp) is the most common of these.

Signal processing in automation systems is not limited to digital signals because some systems are entirely analog and others use both digital and analog signals. The circuits that process analog signals are quite different in their operation. Most of them operate in a linear mode so that all frequencies of the input signal are preserved in the amplified output.

By far, the most common linear circuit is the operational amplifier, often called an op-amp. Its frequency range is between 0 and about 1 megahertz. For higher frequencies a video amplifier must be used, but the need for such a device in automation systems is practically non-existent. Op-amps also are limited in their power output and any need for more than about 5 watts requires the use of power amplifiers. Power amplifiers are built around large discrete transistors which usually must be mounted on a heat sink. The following discussion of linear circuits is limited to integrated circuit op-amps.

The Basic Operational Amplifier

Op amps are high-gain IC amplifiers that have two inputs and one output. One input produces an out-of-phase (inverted) output, the other an in-phase (non-inverted) output.

The circuit symbol for a linear op-amp is shown in *Figure 4-22a*. It has two inputs: one inverts its input at the output (the " − " terminal) and the other does not (the " + " terminal). The open-loop gain of the amplifier is about 100,000, so that theoretically, a 0.1 volt input would produce a 10,000 volt output. However, this is not possible with the power supply voltages available; therefore, the output goes to a limit of about 10 volts (saturates) and the operation becomes nonlinear with only a small input voltage. Because of this, the open-loop configuration is not very useful for linear operation.

Figure 4-22.
The Operational
Amplifier in Three
Configurations

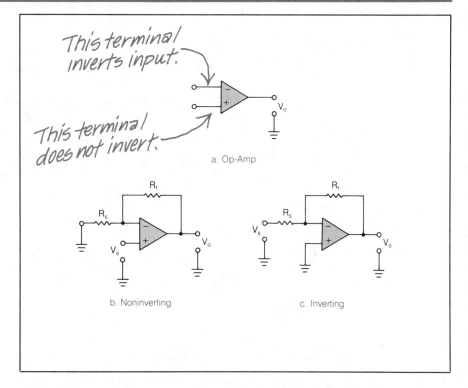

This terminal inverts input.

This terminal does not invert.

a. Op-Amp

b. Noninverting

c. Inverting

More useful arrangements use a closed-loop like those shown in *Figure 4-22b* and *Figure 4-22c*. In (b), the gain of the non-inverting amplifier (the output divided by the input) is:

$$G_{NI} = \frac{V_o}{V_s} = 1 + \frac{R_f}{R_s}$$

and in (c) the gain of the inverting amplifier is:

$$G_I = \frac{V_o}{V_s} = \frac{-R_f}{R_s}$$

The resistance, R_s, is included in the input lead to set the input impedance. The resistor R_f is called a feedback resistor because it forms a small closed-loop system where part of the output signal is fed back to the input to control the gain. These circuit arrangements are widely used as amplifiers for analog signals.

Other Op-Amp Configurations

Three uses of op-amps for automation systems are shown in *Figure 4-23*. A circuit called an adding circuit or summer and its equation is given in *Figure 4-23a*. Notice that the output is a function of the sum of the inputs and is inverted as indicated by the minus sign in the equation. The summer is used in digital to analog converters which will be discussed a little later.

**Figure 4-23.
Three Types of Op-Amps**

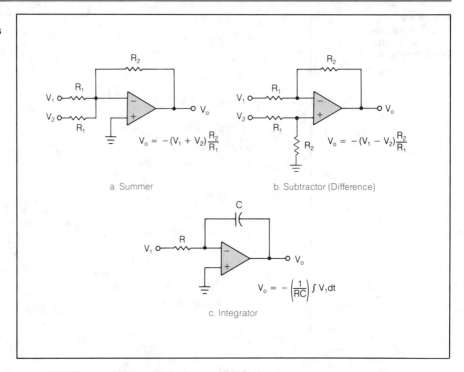

$$V_o = -(V_1 + V_2)\frac{R_2}{R_1}$$

a. Summer

$$V_o = -(V_1 - V_2)\frac{R_2}{R_1}$$

b. Subtractor (Difference)

$$V_o = -\left(\frac{1}{RC}\right)\int V_1 dt$$

c. Integrator

Op amps are used to perform adding, subtracting and integration of analog input signals.

A subtractor and its equation are given in *Figure 4-23b*. The output is the difference of the input signals multiplied by the gain of the amplifier. The subtractor is used in situations where only the difference between two signals (such as desired temperature − actual temperature) is required, not in the absolute values of the signals themselves.

The third circuit is an integrator. As indicated by the integration symbol in its equation in *Figure 4-23c*, an integrator continuously sums the input to generate the output. The most useful application in automation systems involves a constant input where the output of the integrator is the input multiplied by the time that this input has been applied and this product multiplied by the gain. That is, if an input of 2 volts is applied for 0.5 second, the output of the integrator is 1 volt if the gain is one. Notice that the integrator circuit is like a summer except the feedback resistor is replaced with a capacitor. The capacitor performs the integrating function. The integrator circuit is an important element in proportional-integral (PI) and proportional-integral-differential (PID) control systems.

Review

Three general types of electronic building blocks used in automation systems have been discussed: combinational blocks, sequential blocks, and analog blocks. More complex circuits, subsystems and systems use combinations of these building blocks. Two very important ones for control systems are data conversion subsystems for signal conditioning and the microcomputer. Let's discuss each of these.

DATA CONVERSION

Most systems to be controlled are analog. The controllers are digital. Thus, there is a need for D/A and A/D converters.

Digital computers require digital binary signals—those that may have only two possible voltage levels—as inputs, and they generate digital signal outputs. Unfortunately, most systems to be controlled operate as analog systems and many of the sensors generate analog signal inputs and the system outputs analog signals to operate actuators. Consequently, analog to digital (A/D) converters and digital to analog (D/A) converters must be used between the digital computer and the controlled system. This is the signal conditioning that was mentioned earlier. The concept of D/A and A/D conversion is built around a principle called Nyquist's Sampling Theorem:

If an analog signal is uniformly sampled at a rate at least twice its highest frequency content, then the original signal can be reconstructed from the samples.

The theorem can be described by the equation:

$$f_s \geq 2BW$$

The sampling frequency is f_s and the bandwidth of the signal (which defines the highest frequency) is BW.

Sampling an Analog Signal

In A/D conversion, the analog signal must be sampled at twice the highest frequency of the signal being sampled in order to reproduce the signal accurately.

Figure 4-24 shows how an analog signal varies with time. The times at which the waveform is sampled are superimposed on the figure and, as shown, the amplitude of the signal does not change very much in the short time interval between samples. As a result, the sample at the sample time is a close representation of the signal for a short time on either side of the sampling point.

Figure 4-24. Sampling an Analog Signal

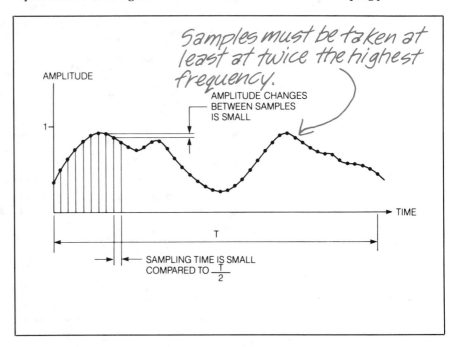

Obviously, more samples in the time T would reproduce the waveform more accurately, but as stated by Nyquist's theorem, if the signal is sampled at a rate which is greater than twice the highest frequency contained in the signal, the samples will contain all the information contained in the original signal. If the maximum frequency of signals that are present in a control system is 100 hertz, then sampling for A/D conversion must be at least 200 times per second.

Analog to Digital

To convert an analog signal to a digital signal, the analog signal is measured at specific intervals, and the value at each interval is converted into a unique digital code.

The basic principles of an A/D converter are shown in *Figure 4-25a*. The input is a continuously varying signal that is sampled at specific times determined by the sampling rate. The sampled voltage value is converted to a unique digital code for that value. In *Figure 4-25a*, an 8-bit code is used. All bits of the code come out of the converter at the same time (in parallel) each time the input is sampled and the code represents the input signal value at the sample time. The codes are sent to other digital circuits in the system in parallel (each bit over a separate wire) or they are sent out one bit after the other in time over a single pair of wires (serially). It should be apparent that if the code is an n-bit code and it is sampled at a rate of s times per second, then the bit rate in bits per second is $n \times s$ when the signal is sent serially down a line.

Figure 4-25.
Signal Conversions
(Source: J. L. Fike and G. E. Friend, Understanding Telephone Electronics, *Texas Instruments Incorporated, Copyright © 1983)*

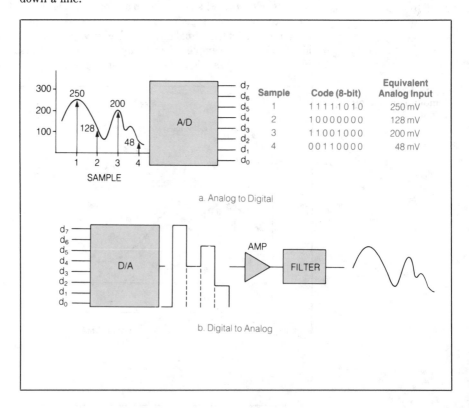

a. Analog to Digital

b. Digital to Analog

Digital to Analog

To convert a digital signal to an analog signal, the stepped output voltages from the converter are passed through an amplifier and filter to recover the original signal.

At the D/A converter, the input is the same or similar parallel code that was output from the A/D converter. (If the code is in serial form, it must be converted to parallel form for use by the D/A converter.) The D/A converter outputs a voltage level corresponding to the input code as shown in *Figure 4-25b*. During the sample period, the voltage level remains constant; therefore, the output has stepped voltage levels. The output is restored very nearly to its original continuously changing shape by passing the signal through an amplifier and filter as shown. The more A/D samples taken, the more closely the filtered D/A signal matches the original.

THE DIGITAL COMPUTER

The heart of the control system is the controller or in many cases today, a microcomputer. In order to be classified as a computer, a system must satisfy five criteria which were estabished in 1830 by Charles Babbage:

Todays microcomputers satisfy the basic requirements for a computer in their four functional units; input, memory, CPU, and output.

1. It must have an *input* capability so that data and instructions can be entered.
2. It must have a *memory* to store data, instructions and results.
3. It must be capable of making *calculations* using arithmetic operations.
4. It must be capable of making *decisions* using logical and relational operations in order to select alternative courses of action using the input data and previous calculations.
5. It must have an *output* capability in order to transmit the results of its operations.

The block diagram of a digital computer in *Figure 4-26* shows the four basic blocks that will satisfy all five criteria: an input block, a memory block, a central processing block (CPU), and an output block.

The input block consists of one or more input channels to feed the CPU. A particular input is selected by its address being placed on the address bus by the CPU. The data are transfered to the CPU via the input data bus. Depending on the design, this bus transfers either 4, 8, 16, or 32 bits in parallel. Each bus consists of a bundle of wires insulated from each other with one wire needed for each bit to be transferred (e.g., simultaneous transfer of 8 bits requires 8 wires).

The memory must be capable of storing binary code received from the CPU (write operation) and of providing the CPU with the binary code stored in a memory location (read operation) upon command from the CPU. The memory location is selected by the CPU using the address bus; the data is transferred via the bi-directional data bus.

The memory can consist of semiconductor ROM or RAM, or magnetic storage or a combination of these. The memory of most microcomputers consists of semiconductor RAM and ROM with magnetic disk used for external mass storage.

**Figure 4-26.
Typical Computer Block
Diagram**

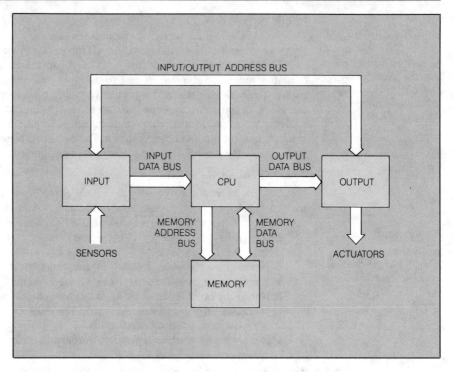

The CPU is the heart of a
computer system.

The heart of the computer is the CPU. It controls operation of the computer, decodes (interprets) and executes instructions, performs arithmetic and logical operations, receives input data and sends output data. It contains many registers, counters, and combinational logic devices. It generates all addresses that are placed on the address bus. It also can perform non-standard tasks based on the presence of input signals called interrupts.

The output block channels binary data from the CPU to the user of the information. The user may be printers, lights, 7-segment display devices, CRT displays, magnetic disk units, and magnetic tape units. Some devices such as magnetic disk units and magnetic tape units are both input and output devices because they are used as external memory for mass storage. Some actuator devices may use the digital output data directly, but others require a D/A converter.

THE MICROPROCESSOR AND MICROCOMPUTER

If the CPU is contained on one integrated circuit, this IC is called a microprocessor. The first microprocessor was developed by the Intel Corporation in 1971.

A microcomputer is a computer whose CPU is a microprocessor. Solid-state technology has developed to the point where it is possible to put the CPU, memory, and input/output (I/O) on one integrated circuit. These devices are called single-chip microcomputers. The first microcomputer for a one-chip calculator was developed by Texas Instruments in 1971 and for a single-chip 4-bit microcomputer in 1972.

Figure 4-27 is a single-chip microcomputer with the building blocks of the computer identified. The CPU consists of the arithmetic and logic unit (ALU) and the control decode section. The ROM has 1,024 locations and each location can store 8 bits of information. The RAM has 64 locations of 4 bits each. The oscillator clock block will be discussed shortly.

**Figure 4-27.
Single-Chip
Microcomputer**
(Source: D. L. Cannon, G. Luecke, Understanding Microprocessors, *Texas Instruments Incorporated, Copyright © 1979)*

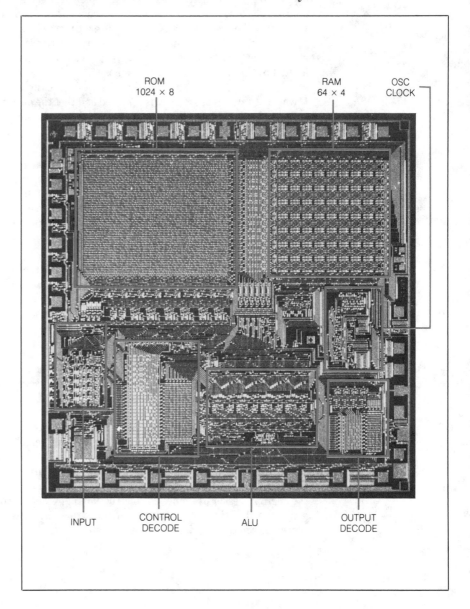

What the Computer Does

By performing fairly simple tasks in a predetermined sequence of steps at tremendous speeds, computers can solve very complex tasks quickly and accurately.

Computers can perform very complex tasks, but they do this in amazingly simple steps. They are very predictable and the secret of their success is the speed with which they operate, not the complexity of their operation. All computers perform a series of operations based on the instructions given to them. The set of instructions is called a program and is very much like a recipe in a cookbook. Executing the instructions in the order given will result in the desired output. The instructions are stored in memory at particular locations in sequence as binary codes and each instruction is located by sending its address to memory.

As an example, assume that the average of three numbers A, B, and C which are stored in memory is required and that the answer should be outputted on a printer. The set of instructions could be something like this:

1. Fetch A from memory and store it in the accumulator register in the CPU.
2. Fetch B from memory, add it to A, and store the result in the accumulator register.
3. Fetch C from memory, add it to the contents of the accumulator register. Store the result in the accumulator register.
4. Divide the sum by 3 and store the quotient in the accumulator register.
5. Output the contents of the accumulator register (the answer) to the printer.

This program would require only a few millionths of a second to execute.

For the purposes of this book, one need not be an expert in microprocessors, but merely needs to understand the basics of what they do and how they fit into electronic control of automation systems.

How the Microprocessor Follows Instructions

The master clock generates preset, controlling output pulses, to accurately time all operations of the system.

The microprocessor operation is governed by a master clock signal that is generated in the clock block of *Figure 4-27*. The clock generates a constant stream of output pulses at a preset rate or frequency. This preset rate is called the clock cycle. Several clock cycles are required to complete the tasks associated with one instruction.

The smallest unit of operation in a microprocessor is called the machine cycle. It consists of two components—the "fetch" cycle and the "execute" cycle. During the fetch cycle, the address of the next instruction is sent to memory and the instruction stored at that address is returned to the microprocessor (i.e., the instruction is "fetched"). During the execution cycle, the microprocessor examines the instruction that was fetched and takes some action based on that instruction (i.e., the instruction is "executed"). This process is repeated until the computer is stopped.

How the Microprocessor Knows Where to get the Next Instruction

Initial instructions are stored at special memory locations to get the computer system started at the same place each time it is turned on. Then it is ready to run other programs.

There is a special register in the microprocessor that is called the program counter. It always contains the address of the next instruction. When the computer is turned on, the computer registers and memory are prepared to accept an external program. This preparation is called initialization. After initialization, the program counter contains the address of the first memory location that can store a user program. When a program is input to memory, as stated previously, each instruction is stored in sequential locations beginning at the first location allocated. Then, when the program is "run" or executed, the program counter is set to fetch the first instruction. After the first instruction is fetched, normally the program counter is advanced to the address of the next sequential instruction and the program continues in that fashion. In many cases, the fetched instruction itself tells the location of the next instruction to be fetched.

Consider again the averaging program that was discussed previously. Assume that when the computer is initialized, the program counter is set to memory location 001. The instructions for the averaging program could be as given in *Figure 4-28*. During the first machine cycle, the microprocessor fetches the instruction at location 001 which is to load A. This is shown pictorially in *Figure 4-29a*. The program counter contains 002 so the value of A (which is 5) is fetched and put in the accumulator as indicated in *Figure 4-29b*. The program counter now contains 003 and the next instruction (Add B) is fetched. This process continues until the instruction at memory location 008 is fetched (*Figure 4-29c*). This instruction changes the contents of the program counter to 200 (*Figure 4-29d*) and execution continues with the next instruction fetch at memory location 200 (*Figure 4-29e*).

**Figure 4-28.
Program for Averaging
Three Numbers**

Programming allows user to change operations without changing machinery.

Memory Location	Instruction	Comments
001	LOAD A	PUT A IN ACCUMULATOR
002	A	DATA
003	ADD B	A + B IN ACCUMULATOR
004	B	DATA
005	ADD C	A + B + C IN ACCUMULATOR
006	C	DATA
007	DIVIDE 3	(A + B + C)/3 IN ACCUMULATOR
008	GO TO 200	NEXT INSTRUCTION AT MEM LOC 200
009		
010		
200	STORE 300	PLACE ANSWER IN MEM LOC 300
201	OUTPUT 2	OUTPUT ANSWER TO DEVICE #2
202	STOP	END OF PROGRAM

**Figure 4-29.
Some Steps in Program
Execution**

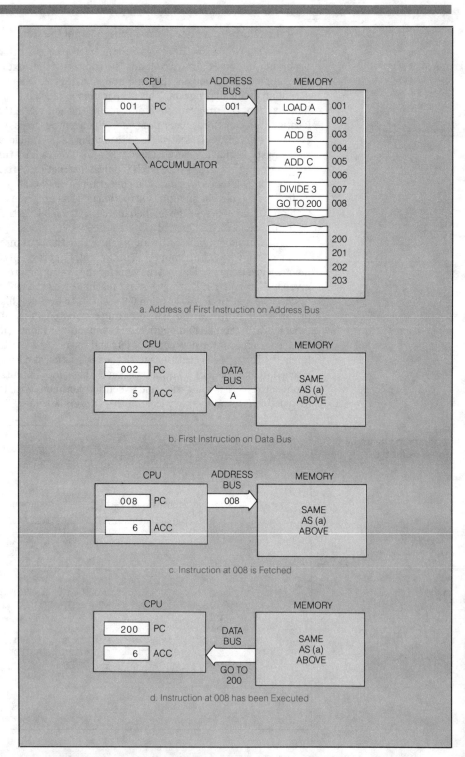

a. Address of First Instruction on Address Bus

b. First Instruction on Data Bus

c. Instruction at 008 is Fetched

d. Instruction at 008 has been Executed

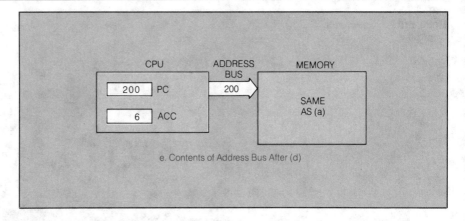

e. Contents of Address Bus After (d)

Usually, the programmer's task is to write or input the instructions in the proper sequence without concern about how they will be stored in memory because this will be done correctly by the computer. The great advantage in the control system is that program steps can be modified, added to, or changed to change the application or the task without changing hardware.

THE PROGRAMMABLE CONTROLLER

The programmable controller (PC) is a type of computer commonly used to automate manufacturing operations. Changes can be made in the system operations by changing the program.

According to the National Electrical Manufacturers Association (NEMA), a programmable controller (PC) is "a digitally operating electronic apparatus which uses a programmable memory for the internal storage of instructions for implementing specific functions such as logic, sequencing, timing, counting, and arithmetic to control through digital or analog input/output modules, various types of machines or processes". PCs have been available for over ten years, but the advent of the microprocessor provided both the incentive and the means for development of a new series of PCs with added capability. Over fifty products on the market meet the NEMA definition of a PC and yearly expenditures on PC systems exceed $200,000,000.00.

The PC has grown exponentially in popularity because it is small, highly reliable, inexpensive and, most importantly, can be programmed by a plant technician or electrician. Upgrading the system is easy to accomplish and troubleshooting is straightforward.

Is The PC a Computer?

If you review the NEMA definition while examining *Figure 4-26*, you will quickly agree that a PC is a computer. It must be a rugged computer to withstand rough handling and to be capable of operating in temperatures from 32° to 140°F (0° to 60°C) 95% relative humidity. It also must be shielded so that it can operate reliably in an electrically noisy environment.

The Texas Instruments 510 PC shown in *Figure 4-30* is a good example of a small PC that can be used to control up to 20 relays, timers and counters. The controller uses a microprocessor and has a 256 location RAM. The handheld keypad shown in the figure is used to program the PC.

**Figure 4-30.
Model 510 Programmable
Controller**

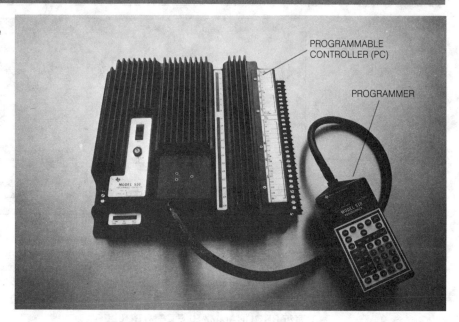

PROGRAMMABLE
CONTROLLER (PC)

PROGRAMMER

The TI510 has twelve inputs that operate at ac or dc levels in groups of six. The eight outputs can be either 120 Vac or 24 Vdc. Since the 510 has a CPU, memory, input, output and is programmable, it is indeed a computer.

Programming the PC

PCs have sets of instructions that are especially designed for industrial applications to make programming easy. The programming is designed to be translated from relay ladder diagrams to keystrokes on the keypad. Easy interconnection, a flexible selection of I/O modules, and an easy interface to industrial power make the PC a very desirable system for many manufacturing control systems. A complete system application will be discussed in Chapter 8.

WHAT HAVE WE LEARNED?

1. The invention of the transistor led to the invention of the integrated circuit which has triggered the digital systems explosion.
2. Building blocks of combinational logic circuits (AND, OR, NOT, NAND, NOR) are combined to make encoders, decoders, and data selectors. These and sequential logic circuits (flip-flops and latches) are combined into high-functional high-density circuits in integrated circuit form that have high performance, low power dissipation, small size, high reliability and yet are low cost.
3. The operational amplifier is an important component used to amplify analog signals in control systems.
4. Analog signals must be converted to digital signals to use as inputs to digital computers. A/D converters do the conversion.
5. Digital codes from digital systems must be converted to analog signals to be able to drive actuators in control systems. D/A converters do the conversion.
6. A computer has input, output, CPU and memory components.
7. Combinational and sequential logic circuits are combined to make a computer CPU (microprocessor) and even a complete microcomputer in integrated circuit form.
8. A programmable controller is an important computer designed specifically for automation systems.

Quiz for Chapter 4

1. The integrated circuit:
   a. led to the development of the transistor.
   b. often is called a "chip".
   c. has revolutionized electronics.
   d. b and c.

2. An encoder:
   a. is used to send telegraph signals.
   b. converts an input to a binary number.
   c. can have only four keys.
   d. converts a binary number to a decimal number.

3. A flip-flop:
   a. is a sequential logic device.
   b. is a combinational logic device.
   c. remembers what was previously stored in it.
   d. a and c.

4. The parallel register:
   a. can store a multiple-bit binary number.
   b. uses R-S flip-flops.
   c. is erased after a word is read from it.
   d. none of the above.

5. An operational amplifier:
   a. can be used to sum two or more signals.
   b. can be used to subtract two or more signals.
   c. uses the principle of feedback.
   d. all of the above.

6. A/D conversion:
   a. changes an analog signal to a sequence of binary numbers.
   b. changes binary numbers of one type to another type.
   c. is sampled at ½ the highest input frequency.
   d. must be fed through a filter.

7. D/A conversion:
   a. results in a binary code as the final output.
   b. has to be sampled at ½ the highest input frequency.
   c. must be fed through a filter to reproduce the original signal accurately.
   d. b and c above.

8. A microprocessor:
   a. is a CPU in integrated circuit form.
   b. is another name for a computer.
   c. executes a program of instructions.
   d. a and c.

9. The program counter:
   a. counts the number of instructions that have been executed.
   b. decides which program to run.
   c. contains the address of the next instruction to be executed.
   d. is where arithmetic is done in a computer.

10. When a computer fetches an instruction, the binary code representing the instruction appears on the:
   a. I/O address bus.
   b. memory data bus.
   c. output data bus.
   d. input data bus.

Software/Programming Languages

ABOUT THIS CHAPTER

The best designed programmable digital electronic process control system in the world is worthless unless it is given instructions for the sequence of operations that it is to perform and for what to do for every conceivable situation that may arise. In this chapter, discussions help to explain the process by which instructions are developed and stored in an automation system, and the kinds of languages available for stating the instructions and describing the operation to be performed.

WHAT IS SOFTWARE?

The operation of a conventional electromechanical machine is determined by how the machine parts work together; thus, the actual position of a gear, or lever, or wheel, or shaft usually reveals the machine's reaction for any movement of one of its parts. *Figure 5-1* shows a simple electromechanical machine. Applying power to the motor by closing the switch causes the fan to turn. The fan speed is a result of motor speed, belt and pulley sizes, and gear ratios.

Conventional machinery is operated by the physical interaction of its components. Skilled observers can often tell how well a machine is operating by watching its components function.

**Figure 5-1.
An Electromechanical
Machine**

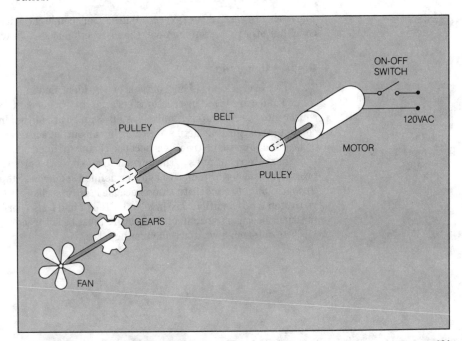

In computer-controlled systems, it is not always possible to look at the way the machine is constructed to determine what the machine will do. A computer has replaced some of the conventional machine parts and the system operation is governed by a set of instructions stored in its memory. The term "software" simply refers to a set of instructions that tells the computer what to do for a given set of circumstances.

Remember that the computer is connected to the controlled system through sensors and actuators. The software causes the computer to energize the actuators in a predetermined step-by-step sequence based on inputs from sensors and on computer calculations made by the program based on system parameters.

It is not always easy to recognize what a computer system is doing because computers work more from a set of instructions stored in their memory called "software."

Example Use of Instructions

As an illustration, consider the cooling system in *Figure 5-2*. Whenever the water temperature exceeds 150°F as measured by the temperature sensor, the cooling fan should be turned on. When the water temperature falls below 145°F, the fan should be turned off. The computer turns the fan on or off by the control signal it sends to a relay (the actuator). The following instructions could accomplish this. (These are not in a computer language; actual computer languages will be discussed later.)

Step	Instruction
1	Read the temperature from the temperature sensor.
2	If the temperature is less than 150°F, go to Step 1.
3	Turn on the cooling fan.
4	Read the temperature from the temperature sensor.
5	If the temperature is greater than 145°F, go to step 4.
6	Turn off the cooling fan.
7	Go to Step 1.

The steps are carried out (executed) in sequence unless the instruction tells the computer otherwise; for example, the instruction in step 7 says to go back to step 1.

Kinds of Instructions

Five different kinds of instructions used in controlling the cooling system are: input, output, conditional branch, unconditional branch, and arithmetic.

Several kinds of instructions can be identified in the seven steps. Steps 1 and 4 are *input* instructions because they bring data from an outside sensor into the computer. Similarly, steps 3 and 6 are *output* instructions; the computer energizes an actuator producing an output action. Steps 2 and 5 are *conditional branch* instructions because certain defined conditions determine if the next step continues in the normal sequence of the program or if the program branches or jumps to another defined sequence. Step 7 is called an *unconditional branch* instruction because it *always* changes the sequence to go to step 1; no conditions are involved. Steps 2 and 5 also perform *arithmetic* functions because the actual temperature must be subtracted from 150°F or 145°F to determine what action must be taken.

**Figure 5-2.
Computer Controlled
Cooling System**

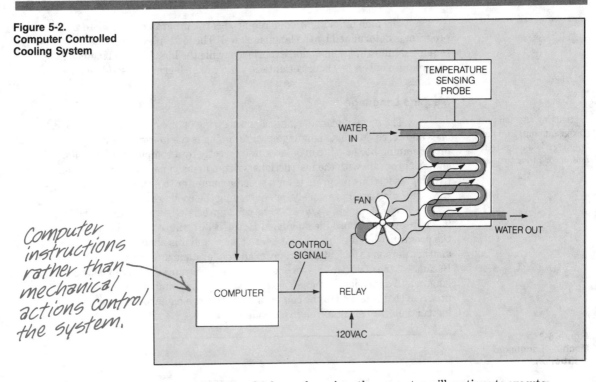

Computer instructions rather than mechanical actions control the system.

Steps 1 and 2 form a *loop* since the computer will continue to execute these two steps until the temperature is equal to or greater than 150°F. Steps 4 and 5 also form a loop and step 7 makes the entire sequence into a third loop.

The *form* or *syntax* of the software instruction depends on the computer that will execute the instructions and the structure of the language used. The set of rules governing the form used to write instructions is called the *program language*. The various program languages are given names and are classified into levels for reference.

LANGUAGE LEVELS

Every computer performs its operations by means of sets of predefined, binary codes called machine language instructions.

Recall from previous discussions that computers operate only with binary numbers; therefore, in its final form, any instruction to a computer must be in the form of ones and zeros and is called a binary coded or machine language instruction. The computer then interprets the binary coded instruction and performs the operation specified by the instruction. For instance, the instruction 10110100 might cause the computer to add two numbers.

All computers have a set of predefined binary codes, called the instruction set, that they can interpret and execute. Microcomputers may have an instruction set of only 70 to 80 instructions whereas large computers may have over 200. Unfortunately, each model of computer has a different instruction set. The Digital Equipment Corporation VAX 11/780 has an instruction set that is not in any way related to the instruction set for a Texas Instruments 990/12.

The amount of knowledge required of the programmer to write a program is determined by the language level. The lowest level languages require significant training and experience while the highest level languages may require only a few hours orientation and can be learned by self-teaching methods.

Machine Language

Instructing the computer (programming) in machine language is a long, tedious and error prone process.

The form of machine language is a string of binary digits consisting of ones and zeros. Machine language programming is the lowest level of programming because it can be executed directly by a computer. *Figure 5-3* shows a program with the instructions written in machine language. Not very exciting, is it? This program is for a specific computer to perform steps 1 through 7 discussed previously for controlling the cooling fan.

By examining *Figure 5-3* one can appreciate how difficult and boring it is to program in machine language. It also is difficult to try to find errors in the program (debugging) when it doesn't work and it is almost impossible for another person to look at the program and understand anything about it. Because of these difficulties, one almost never programs directly in machine language. Higher-level languages are used for programming and a special program that is stored in the computer converts the higher-level language instructions into machine language code.

**Figure 5-3.
Machine Language
Instructions**

```
10110110
00000001
00101100
10000000
10010110
00101101
00100000
10001100
11111111
10110111
11001111
01001010
10110110
00000001
00101100
10000000
10010001
00101110
00110010
10000110
00000000
10110111
11001111
01001010
01111110
00000000
00100000
```

Assembly Language

Assembly language, based on symbolic words called mnemonics which closely resembles the computer operation, is more people oriented than machine language and is correspondingly easier to use.

The next higher-level language is called assembly language. The assembly language program in *Figure 5-4* corresponds to the machine language program in *Figure 5-3*. In assembly language, the programmer uses symbol words, called mnemonics, chosen to closely resemble the operation that the computer is to do. Each mnemonic normally corresponds to one computer operation, but this may be translated into more than one machine language instruction. For example, compare the number of machine language instructions in *Figure 5-3* to the number of assembly language instructions in *Figure 5-4*.

**Figure 5-4.
Assembly Language
Instructions**

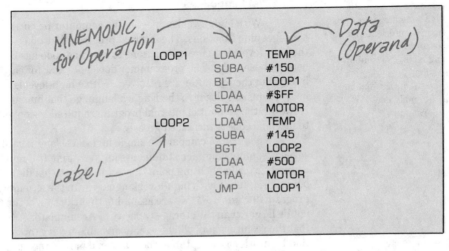

Assembly language instructions have two mnemonic code segments; the first is the basic instruction, the second is the operand which the computer will act upon per the basic instruction.

Any of the mnemonic instructions may have a second part that modifies the command mnemonic by providing additional information (data) necessary for the computer to execute the instruction. Consider the first instruction:

    LDAA TEMP

This instruction means to LOAD accumulator register A with the value stored in a memory location named TEMP. Interpretation of the other commands in LOOP1 are:

SUBA #150	Subtract 150 from the temperature value stored in accumulator A and put the result in A.
BLT LOOP1	If the accumulator value is less than zero, go to the instruction labeled LOOP1; otherwise
LDAA #$FF	Load A with a specific number (which will turn on the motor).
STAA MOTOR	Store contents of accumulator A in a location called MOTOR (turns on the fan motor).

The commands in LOOP2 are the same except these two:

BGT LOOP2	If the accumulator value is greater than zero, go to the instruction labled LOOP2; otherwise
JMP LOOP1	Jump (go) to the instruction labeled LOOP1.

The BLT and BGT are conditional branch instructions and the JMP is an unconditional branch instruction. It probably is apparent that assembly language is much easier to use than machine language since the mnemonics are easy to memorize and understand. However, it is still tedious to program using assembly language because the programmer must pay attention to every detail regarding register contents, overflow/underflow, memory locations, etc. In fact, the programmer must manage all the resources of the computer including allocation of computer memory. (These operations are collectively called "housekeeping".) Assembly language also is machine dependent; that is, each model of computer has its own set of mnemonics.

High-Level Languages

High level languages are the most efficient and the closest to the English language. The computer handles all details rather than the programmer, as in assembly language programming.

With high-level languages, the computer performs its own housekeeping functions. The internal program that takes care of converting the high-level language instructions to machine code has been designed to handle these tasks. The programmer does not know (or care) how the problems internal to the computer are resolved. With a high-level language, the instructions are closer to the English language that humans understand. However, the amount of internal program required to convert the instructions to machine code is increased greatly.

Figure 5-5 compares a single high-level language statement with the seven assembly language statements for averaging five numbers. Decimal arithmetic (called floating point) is very tedious to handle with assembly language, but not with high-level languages. In the assembly language program in *Figure 5-5*, we are assuming that the sum of the five numbers will not be larger than the storage capacity of Accumulator A (no overflow) and that the resulting sum will be evenly divisible by five (no fractional quotient) or that truncation is acceptable (the fractional part of the quotient is dropped without rounding). Otherwise, floating point arithmetic would have to be used.

**Figure 5-5.
Comparison of High-level and Assembly Language Statements**

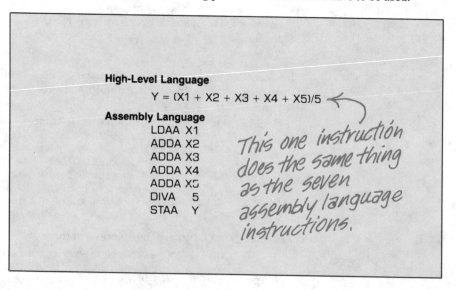

High-Level Language

$$Y = (X1 + X2 + X3 + X4 + X5)/5$$

Assembly Language
```
LDAA  X1
ADDA  X2
ADDA  X3
ADDA  X4
ADDA  X5
DIVA  5
STAA  Y
```

This one instruction does the same thing as the seven assembly language instructions.

High-level languages are called "procedure-oriented" languages because *they allow programmers to concentrate on the problem to be solved* rather than the details of checking data values, lengths and movement. A further advantage of high-level languages is that they are less likely to depend on the type of machine; thus, a program written in a high-level language sometimes can be used on another type of machine with little or no modification.

Many high-level languages exist and a few are listed by name in *Figure 5-6*. Some are "higher" than others and require the user to have very little knowledge of programming or of the computer on which the program will be run. These are often "menu-driven" where the user simply selects from a list of alternative items presented on a screen in English language and a special computer program "writes" an application program for the task.

The newest of the menu-driven languages are picture coded rather than word coded. The "programmer" simply moves a pointer (cursor) to the picture that represents the necessary action by using a joystick or similar device. Once the picture is chosen, a set of instructions predefined by the manufacturer is executed. Pictures are selected in sequence to indicate the step-by-step action that is to be executed.

The goal of procedure-oriented languages is to make the power of the computer available to many people with a minimum of training. The design of equipment to control automation systems will be greatly influenced by the development of new procedure-oriented high-level languages over the next decade. However, if a person is going to write a program, that person must know some things about the structure of the program—what kinds of things the language that it is written in can do. That's what we'll discuss next.

With high-level language programming, the programmer can concentrate on problem solving and analysis and the program will likely run on several computers.

**Figure 5-6.
Some High-Level
Languages**

General Purpose		Special Purpose	
FORTRAN	Scientific	APT	Machine Tools
ALGOL	Scientific	ADAPT	Machine Tools
PASCAL	Scientific	GASP	System Simulation
BASIC	Scientific & Business	CSMP	System Simulation
PL/1	Scientific & Business	ECAP	Circuit Analysis
COBOL	Business	SPICE	Circuit Analysis

WRITING INSTRUCTIONS

The language instructions are divided into segments called "fields", such as, label, command, operand and comments.

Suppose that two numbers are to subtracted. How are the instructions written in the language that's being used? All languages have rules about how the instructions must be written. The various components of an instruction are called fields. The *label* field allows identification of the instruction for later reference. In *Figure 5-4*, LOOP1 and LOOP2 are in label fields and allow branching to those instructions from other instructions.

A *command* field defines the instruction to be executed by the computer. In *Figure 5-4*, LDAA, SUBA, BLT, etc. are in command fields. In the high-level language statement of *Figure 5-5*, the expression for Y is a command field.

Medium- and low-level languages also have an *operand* field. This field allows modification of the command in the command field or to supply information needed to carry out the command.

Finally, many languages also have a *comment* field. Comments are ignored by the computer. They are included for humans who need to understand the program in order to use it or to find errors if the program does not execute correctly.

In the instruction:

HOGS CPR, A,B   Compare A and B

HOGS is the label; CPR is the command for compare; A,B are the variables to be compared; and Compare A and B is a comment.

For the instruction:

247 DO 250 J = 1,10

247 is the instruction label; DO 250 is the command; J = 1, 10 is the operand. There is no comment field in this statement.

Input/Output

The input/output section provides the means to communicate with the computer and the computer to interface with other equipment.

A programming language must provide for instructions that allow the computer to communicate with the sensors, actuators and other pieces of equipment connected to it because programs must have instructions to input and output data to/from the system and to communicate with the operator. Consider:

500 INPUT "WHAT IS THE MAX SPEED"; S

This statement causes the message enclosed in quotation marks to be printed on a CRT or printer to ask the operator to input the maximum speed. The value typed by the operator will be assigned to S and stored in memory in a specific location so the computer can find it later.

Decisions

The computer's instruc-
tions also allows logic decisions to be made (based on
conditional inputs and alternate choices) and mathematical calculations to be
performed.

Logical conclusions are drawn and alternate paths are chosen based
on values of variables in the system that is being controlled. These conclusions
are usually stated in IF statements such as

IF X<Y THEN MAX: = Y
ELSE MAX: = X;

Boolean logic involving AND, OR, NOT, etc. also may be used in determining
choices. For example, the instruction,

IF A = 5.0 AND B = 8.2 OR C = 5.77 THEN GO TO HOGS

will cause the program to branch to and execute the instruction identified by
the label HOGS when either the condition A = 5.0 and B = 8.2 is true, or if
the condition C = 5.77 is true; otherwise, the next instruction in the normal
program sequence is executed.

Mathematical Calculations

Many control functions require the controller to perform complex
mathematical operations. This requirement is accommodated easily by most
languages since they provide for the calculation of routine arithmetic and
trigonometric functions. Special mathematical routines such as integration and
curve fitting also are available as a single command in some languages.

EXAMPLES OF PROGRAMMING LANGUAGES

Examining programming
languages that are used
for automation computer
control gives us an appreciation of their style and
their ease of use.

Three programming languages suitable for use with computer control
of automation systems will be briefly described. These are:
1. The language used to control the TI510 Programmable Controller.
2. APT—the language that can be used to program numerically
   controlled machine tools.
3. FORTRAN—the language that can control large complex systems that
   are either continuous or discrete in nature.
It is not the intent to present sufficient detail for one to be able to program in
any of these languages, but to give an appreciation for the style of the
languages and the ease with which they may be used.

TI510 PC Programming Language

Using the TI510 language,
a drum timer function is
formed. Each one of the
drum timer input steps
will cause the drum timer
to perform a different
task. The time between
steps can be changed by
the programmer.

The TI510 programming language is used to allow computer control of
outputs based on input values. The input-output relationship is described by
means of ladder logic.

The only elements allowed are input, output and control relay
elements and counter and timer elements. The symbols for these are shown in
a typical ladder diagram in *Figure 5-7*. The ladder diagrams will be explained
in Chapter 8.

**Figure 5-7.
A Ladder Logic Diagram**

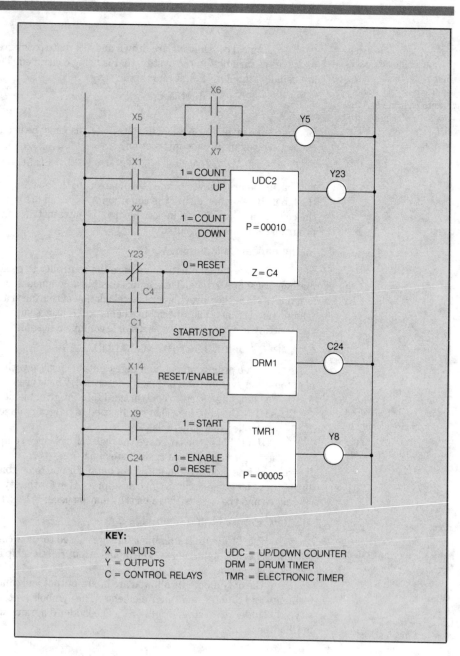

KEY:

X = INPUTS
Y = OUTPUTS
C = CONTROL RELAYS

UDC = UP/DOWN COUNTER
DRM = DRUM TIMER
TMR = ELECTRONIC TIMER

The drum timer is a 50-word instruction with a series of timed outputs (up to 15) which are turned on sequentially as the drum executes 16 steps. The time between steps is under programmer control and allows execution of several operations in sequence. The input C1 starts and stops the drum timer, but does not reset it. The X14 input is the reset/enable line. If X14 is closed, the drum is enabled and can be started and stopped using C1. The drum moves to the next step when all conditions in the current step have been satisfied. C24 is an external output that is enabled when the timer has completed its sequence of operations.

The counter shown is an up/down counter. (Up-only and down-only counters also are available with only a change in the inputs necessary.) The bottom input to the counter resets or enables it. If the input is open, the counter is disabled and reset. The top input causes the counter to count up each time X1 is closed. Likewise, the middle input causes the counter to count down each time X2 is closed. The output Y23 is enabled to reset the counter when the counter has counted up to its preset value or down to zero. P is the preset count (10 in this example) and Z is an output device that is also enabled when the counter reaches zero.

The other timer (TMR1) is started when X9 is closed if C24 also is closed. The timer decrements from a preset value (maximum is 32,767) to zero with one decrement occurring every 16.67 milliseconds. In this example, the preset value is 5 so 16.67 x 5 or 83.35 milliseconds after the timer is started, the output, Y8, is enabled. Opening C24 stops the timer at any time and resets the timer to its preset values. Closing C24 enables the timer for use.

Basic Instruction Set

The instruction set for this language consists of only 5 instructions:

Mnemonic	Action
STR	Store
OUT	Output
AND	Series Components
OR	Parallel Components
NOT	Reverse Action

The STR instruction is used to indicate the start of each new rung in the ladder diagram.

Example Program

Let's see how we use these instructions to describe the system of *Figure 5-7*. The first rung is straightforward and may be written as:

```
STR X5
AND X7
OR X6
OUT Y5
```

The up/down counter is programmed by specifying each of its three input lines, then specifying the counter itself. Next the count value and the element to be enabled when the counter is zero are given. For the UDC 2 counter in *Figure 5-7*, the program would be:

```
STR X1
STR X2
STR NOT Y23
OR C4
UDC 2
10
C4
OUT Y23
```

The timer is similar. It's program would be:

```
STR X9
STR C24
TMR 1
5
OUT Y8
```

The drum timer is straightforward and approached in the same way; therefore, a discussion of it would not add further to the understanding of this type of programming language. Its programming is quite detailed and is not included here due to limited space.

AUTOMATICALLY PROGRAMMED TOOLS (APT) LANGUAGE

APT (automatically pro-
grammed tool) tool motion
uses a coordinate system
to describe how a piece is
to be machined.

APT is widely used in the United States to control positioning, continuous-path and contour movements of a machine tool. It is a high-level language with several hundred instructions in its instruction set. These instructions can be grouped into four categories: Geometry, Motion, Postprocessor, and Auxiliary.

Geometry Instructions

It is necessary to describe the location of points and surfaces on a workpiece so that tool motion can be defined and specified relative to these points and surfaces. APT assumes a coordinate system as shown in *Figure 5-8*. Translational motion is described in terms of dimensions in the X, Y and Z directions and rotational motion in degrees about the X, Y, Z axes. The discussion here will be limited to translational motion and the work table is assumed to be in the X-Y plane.

An APT geometry statement has the following form:

symbol = type of geometry/description

For instance, to specify a point:

    P1 = POINT/1.7, −3.4, −2.5

where 1.7, −3.4, and −2.5 are the X, Y, Z coordinates of the point designated as P1. (The coordinates must be given in the X, Y, Z order.) P1 is plotted on *Figure 5-8.*

Figure 5-8.
Coordinate Systems for
the APT Language

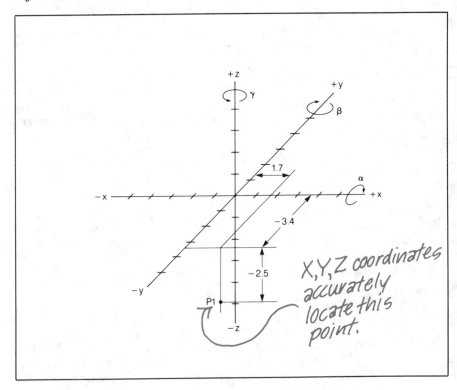

By changing the instructions and geometric coordinates, tools operate at discrete points, in circular patterns, or along variable lines or planes.

Also, a point may be specified as the intersection of two previously defined lines:

    P1 = POINT/INTOF, L5, L7

Here L5 and L7 are lines that have been previously defined. For example, L5 could be defined by:

    L5 = LINE/P2, P3

where P2 and P3 are previously defined points.

To define a circle, the circle must be in the X-Y plane and can be described in a number of ways. Two ways are:

    CIRC1 = CIRCLE/CENTER, PNT, RADIUS, 3.7
    CIRC2 = CIRCLE/P1, P2, P3

CIRCle 1 is described by locating the center at a previously defined point, PNT, and giving the radius as 3.7. CIRCle 2 is described by the previously defined points P1, P2, P3 on the circumference of the circle.

A plane can be specified as:

```
PLN1 = PLANE/P1, P2, P3
```

where the plane passes through the three previously defined points. Another useful way of describing a plane is to refer to a previously defined plane. For example:

```
PLN2 = PLANE/P1, PARLEL, PLN1
```

defines a plane, PLN2, that passes through the point P1 and is parallel to the previously defined plane, PLN1.

Example

Now let's use these instructions to describe the workpiece shown in *Figure 5-9*. The perimeter of the part, the upper and lower surfaces, and the location of the hole must be described. An APT geometry description might be:

```
P1  = POINT/4.0, 0.0, 0.0
P2  = POINT/4.0, 4.0, 0.0
P3  = POINT/3.0, 2.0, 0.0
P4  = POINT/1.0, 3.0, 0.0
P5  = POINT/1.0, 1.0, 0.0
L2  = LINE/P1, P2
C1  = CIRCLE/CENTER, P4, RADIUS, 1.0
C2  = CIRCLE/CENTER, P5, RADIUS, 1.0
C3  = CIRCLE/CENTER, P3, RADIUS, 0.25
L1  = LINE/P1, LEFT, TANTO, C2
L4  = LINE/LEFT, TANTO, C2, LEFT, TANTO, C1
L3  = LINE/P2, RIGHT, TANTO, C1
PL1 = PLANE/P1, P2, P4
PL2 = PLANE/0.0, 0.0, −0.75
```

The statements describing L1, L3 and L4 may need some explanation. L1 is a line passing through P1 and tangent to (TANTO) the left side of circle C2 as viewed from P1. L4 is tangent to C2 and C1 on the left side as viewed from the first circle mentioned in the instruction, which in this case is C2.

Motion Instructions

Motion instructions con-
trol the tool between coordinate points to machine
straight lines or drill
holes.

Motion instructions are used to describe the path of movement of the machine tool. The form of a motion instruction is:

```
MOTION COMMAND/DESCRIPTION
```

The command indicates what to do and the description indicates where to do it. The two kinds of motion are point-to-point (PTP) and contouring (continuous motion). Before either type is used, a starting point (initial location of the tool or reference) must be given with the FROM statement. For instance:

```
FROM/2.0, 3.0, 2.0
```

defines that the tool is initially located 2 units to the right, 3 units behind and 2 units above the origin of the workpiece reference axes.

Figure 5-9.
A Workpiece

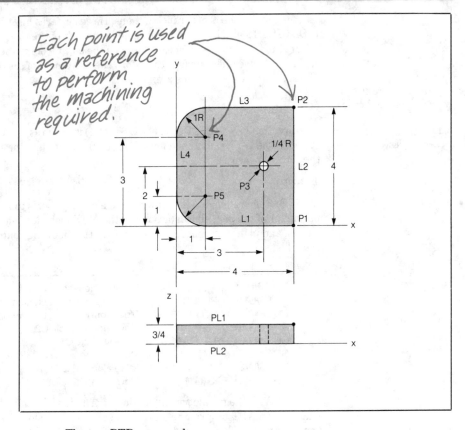

Each point is used as a reference to perform the machining required.

The two PTP commands are:

GOTO moves the tool to the point described
GODLTA moves the tool the described distance from the last tool location

If the tool is currently located at (1.0, 2.0, 0.0) and it is to be moved to (2.5, 3.0, 1.0), the instruction would be:

GOTO/2.5, 3.0, 1.0

or

GODLTA/1.5, 1.0, 1.0

Refer again to *Figure 5-9.* If the drill bit center home or "at rest" position is located at the coordinates (0, 0, 1), then the following instructions will cause the hole centered at P3 to be drilled (assuming the proper bit is installed):

Instruction	Comment
FROM/0.0, 0.0, 1.0	Initial Position
GOTO/P3	Move drill bit to point 3.
GODLTA/0.0, 0.0, −1.25	Feeds the bit through the workpiece.
GODLTA/0.0, 0.0, 1.25	Retracts the bit.
GOTO/0.0, 0.0, 1.0	Returns to home.

To use the contour instructions, three surfaces must be defined:
1. Drive Surface: The surface that guides the *side* of the tool.
2. Part Surface: The surface that guides the *bottom* of the tool.
3. Check Surface: The surface that *stops* the movement of the tool.

The six contour instructions are:

Contour machining instructions require that the three possible surfaces to be machined be defined and programmed in the proper sequence.

GOFWD	GOLFT	GOUP
GOBACK	GORGT	GODOWN

The direction given (UP, DOWN, LEFT, etc.) is with respect to the last check surface. The GOTO command must be used first to define the starting point of the three surfaces.

Example

Referring again to *Figure 5-9*, if a cutting tool is to follow the perimeter of the part shown (assuming the previously defined point, lines, circles, and planes are part of the program), the following contour instructions could be used:

Instruction	Comment
FROM/2.0, 3.0, 2.0	Initial tool location.
GOTO/L1, TO, PL2, TO, L2	Starting point for machining (P1).
GOLFT/L1, TANTO, C2	Move from P1 along L1 until C2 is reached.
GOFWD/C2, PAST, L4	Move around circle C2.
GOFWD/L4, TANTO, C1	Move to circle C1.
GOFWD/C1, PAST, L3	Move around circle C1.
GOFWD/L3, PAST, L2	Move along L3 to L2.
GORGT/L2, PAST, L1	Move along L2 to L1.

With the geometry and motion instructions described and shown by example, an APT program to move a tool to a point or along a surface of any three-dimensional part could be written. The remaining two categories of APT instructions control the machine tool itself and provide additional programming flexibility.

Postprocessor Instructions

It is important to be able to control the machine tool in terms of its speed, feedrate, energization, etc. This category of instructions does just that. Some examples of commands available to the programmer are:

Command	Description
MACHIN	Specifies the machine tool to be used.
FEDRAT	Specifies the feed rate for machine tool in inches per minute.
COOLNT	Turns coolant on or off.

Auxiliary Instructions

These instructions are a miscellaneous collection of necessary or optional commands useful to the programmer. Some examples are:

CUTTER	Defines the diameter of the cutting tool.
FINI	Indicates the end of all operations.
PARTNO	Defines the part number of the piece to be processed.

FORTRAN—A COMPLEX HIGH-LEVEL LANGUAGE

FORTRAN is a mathematically based, high-level language that has much broader applications than the TI510 or APT language. It is used to manipulate numerical data in the solving of problems, in analysis and for decisions-making.

The first two high-level languages that were explored are rather limited in their application. Nevertheless, they represent important classes of languages likely to be encountered by a person programming automation systems. FORTRAN is a language with more general application. It is a mathematically oriented language and is used primarily for decision-making and analysis based on manipulation of numerical data. Its name comes from FORmula TRANslator.

FORTRAN will be introduced primarily by example since most texts dedicated to teaching the language exceed 100 pages. Insight can be gained simply by examining the arithmetic, control and input/output types of statements.

Arithmetic Statement

The basic arithmetic operations and some of the mathematical functions available are listed in *Figure 5-10*. These can be combined into one statement such as:

$$X = A**1 + SIN((B*C)/(2. - D)) - SQRT(4./Y)$$

**Figure 5-10.
Some FORTRAN
Mathematical Functions**

Function	Example	Explanation
+	C = A + B	Addition
−	C = A − B	Subtraction
*	C = A * B	Multiplication
/	C = A/B	Division
**	C = A ** 3	Exponentiation (A^3)
LOG10	C = LOG10 (A)	Common Logarithm
EXP	Y = EXP (B)	Exponential
SQRT	X = SQRT (D)	Square Root
ATAN	Y = ATAN (X)	ArcTangent
SIN	Z = SIN (B)	Sine
ABS	C = ABS (X)	Absolute Value
MAX	Y = MAX(X1,X2,X3)	Maximum Value

The heirarchy (order) for evaluation is:
1. Expressions inside parentheses.
2. Exponentiation.
3. Functions such as SIN and SQRT.
4. Multiplication and division.
5. Addition and subtraction.

Control Statement

To allow the programmer to alter the normal sequence of operations FORTRAN uses control statements. These control statements transfer control to another process point in the program or memory.

Control statements are those that alter or control the normal sequence of execution of program statements.

GO TO Statements

These statements transfer control, either conditionally or unconditionally to another point in the program. An example of an unconditional GO TO is:

```
GO TO 500
```

where 500 must be the label of an instruction in the program. That instruction is executed next. A modification of the GO TO is:

```
GO TO (10, 15, 18, 35, 81), J
```

Control is transferred to one of the labels within the parentheses based on the value of J. For example, if J is 3, control is transferred to the statement labeled 18.

An example of a conditional GO TO is:

```
IF (X**2 - 5.2) 10, 20, 30
```

The expression $X^2 - 5.2$ is evaluated. (Of course, X must have been assigned a value prior to this statement.) If the result is less than zero, the program goes to the instruction labeled 10; if equal to zero, to 20; if greater than zero, to 30.

DO Loop

The DO loop is another type of GO TO statement. The form is:

```
10 DO 30 I = 1, 100
```

All instructions with labels 10 through 30, inclusive, will be evaluated with I = 1, then with I = 2, etc. until I = 101. At that time, control is passed to the first instruction following label 30.

Input/Output Statements

Input/Output statements allow the computer to communicate with other equipment as well as to receive inputs from the programmer. For FORTRAN these statements are usually READ or WRITE.

These statements are used to regulate the flow of data from input devices (keyboards, tapes, disks, card readers, etc.) to the computer and also from the computer to output devices (printers, tapes, disks, and CRTs). The two primary statements are READ and WRITE. For example:

    READ (5, 8) A, B, N

causes the computer to read data for A, B and N from whatever device is assigned as logical unit 5, and is read according to a format given in the instruction labeled 8. WRITE statements have an identical form. A format statement for the READ statement could be:

    8 FORMAT (F2.2, F5.1, I3)

The numbers in parentheses define the read format of data A, B and N, respectively. The F indicates floating point and I indicates integer. The format numbers define the number of places in the number to be read. For example, F2.2 means the number to be read is floating point with two places on both sides of the decimal point. I3 means the number is an integer with up to three places.

An Example Program

In a certain manufacturing process, five pressure values have been measured during the course of the process and recorded on magnetic tape in the correct format. The system is to analyze the data by finding the maximum pressure and calculating the average pressure. If the difference between any two adjacent data points is greater than 8.1, the data are erroneous; therefore, calculations should be terminated and an error message printed to alert the operator. Assume that the input tape drive is logical unit 2 and the output printer is logical unit 4. A FORTRAN program for doing this is given in *Figure 5-11.*

**Figure 5-11.
Example FORTRAN
Program**

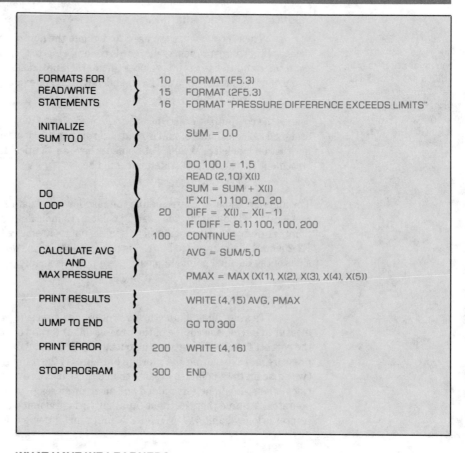

```
FORMATS FOR    }  10   FORMAT (F5.3)
READ/WRITE        15   FORMAT (2F5.3)
STATEMENTS        16   FORMAT "PRESSURE DIFFERENCE EXCEEDS LIMITS"

INITIALIZE     }       SUM = 0.0
SUM TO 0

                       DO 100 I = 1,5
                       READ (2,10) X(I)
DO                     SUM = SUM + X(I)
LOOP                   IF X(I – 1) 100, 20, 20
               20      DIFF =  X(I) – X(I – 1)
                       IF (DIFF – 8.1) 100, 100, 200
               100     CONTINUE

CALCULATE AVG  }       AVG = SUM/5.0
     AND
MAX PRESSURE           PMAX = MAX (X(1), X(2), X(3), X(4), X(5))

PRINT RESULTS  }       WRITE (4,15) AVG, PMAX

JUMP TO END    }       GO TO 300

PRINT ERROR    }  200  WRITE (4,16)

STOP PROGRAM   }  300  END
```

WHAT HAVE WE LEARNED?

1. Software is a set of instructions that controls the operation of an automation system.
2. The form in which the instructions are written is dependent on the computer that will execute the instructions and the language chosen to express the instructions.
3. Languages are classified according to their ease of use.
4. Low-level languages are very efficient, but require knowledge of the computer on which it will be run and a significant amount of training.
5. High-level languages are much easier to use. They may be machine dependent such as the TI510 PC instruction set or quite versatile such as the APT and FORTRAN languages.

Quiz for Chapter 5

1. Software is:
 a. printer output.
 b. a set of instructions.
 c. a programming language.
 d. tape input.

2. Which of the following are kinds of instructions?
 a. Input.
 b. Arithmetic.
 c. Conditional branching.
 d. All of the above.

3. A program language:
 a. defines the form of the instruction.
 b. is always machine dependent.
 c. is never machine dependent.
 d. none of the above.

4. An assembly language instruction:
 a. is written in ones and zeros.
 b. corresponds to one computer operation.
 c. does not require knowledge of the computer design.
 d. none of the above.

5. Which of the following is considered a high-level language?
 a. Machine language.
 b. FORTRAN.
 c. Assembly language.
 d. All of the above.

6. High-level languages:
 a. are "procedure-oriented".
 b. are very difficult to use.
 c. require knowledge of the computer design.
 d. cannot be used with automatic systems.

7. Match the following:
 a. Label field   1. Identifies a specific instruction.
 b. Command field 2. Explains what the instruction does.
 c. Operand field   3. Defines the operation to be performed.
 d. Comment field 4. Gives more information for the command field.

8. Which of the following elements can be used with a TI510 PC?
 a. Drum timer.
 b. Cymbal timer.
 c. Metronome.
 d. Flashing light.

9. The APT language is used with:
 a. large automation systems.
 b. programmable controllers.
 c. drafting systems.
 d. numerical controlled machines.

10. FORTRAN is:
 a. a medium-level language.
 b. machine dependent.
 c. a mathematically oriented language.
 d. all of the above.

11. Which of the following is a machine language instruction?
 a. ADD 5
 b. 10001101
 c. GOTO XOUT
 d. None of the above.

12. A machine tool located at $(4.0, -2.0, 3.5)$ is to be moved to $(3.0, 3.5, 1.0)$ with a GODLTA instruction. Choose the correct instruction:
 a. GODLTA $(1.0, -5.5, 2.5)$
 b. GOTO $(-1.0, 3.5, 1.0)$
 c. GODLTA $(-1.0, 5.5, -2.5)$
 d. GODLTA $(1.0, -5.5, 4.5)$

13. Which of the following are contour
instructions in the APT language?
a. GOLFT
b. GOTO
c. PLN1
d. L3

14. An up/down counter used in a
TI510PC has three inputs. Which of
the following is not one of them?
a. Count up.
b. Count out.
c. Count down.
d. Reset.

15. If the present value in a timer in the
TI510 is 8, how many milliseconds is
required from the time the timer is
started until the output is enabled?
a. 8
b. 133.36
c. 49.88
d. 275.41

16. Which of the following represents a
conditional branching instruction?
a. BGT LOOP5
b. GOTO 50
c. X = 5 + Y
d. JMP MOTOR

17. Which of the follwing is a
FORTRAN instruction?
a. STR NOT Y15
b. GORGT L5, PAST, L3
c. X = R − T/5.0
d. 10101111

18. All computers use the same set of
binary codes.
a. true
b. false

19. In the instruction, 500 SUBA #50
the command field is:
a. 500
b. #50
c. SUBA
d. b and c.

20. The easiest language to use to write
a long, complex program would be:
a. FORTRAN.
b. assembly language.
c. point-to-point
d. machine language.

Continuous Process Control

ABOUT THIS CHAPTER

Chapter 2 described the different types of control systems that are available. This chapter is a continuation of that discussion with the primary focus on how continuous control systems are designed. System applications will be discussed and examples will be given of the design of continuous control systems using a programmable controller and a microcomputer system.

CONTINUOUS PROCESS DEFINITIONS

The definitions for the following terms which are commonly used in process control applications are given in the glossary.

Process Variable	Setpoint
Controlled Variable	System Lag
Manipulated Variable	Transfer Lag
Disturbance	Dead Time
Load Variable	Analog Control
Continuous Control	Digital Control

The example of a hot water temperature control system in *Figure 6-1* will be used to relate some of these terms to a continuous process control system. In *Figure 6-1*, the *controlled variable* is the water temperature, the *load variable* is the demand for hot water, the *setpoint* is the desired water temperature set by the operator, the *disturbance* is the addition of uncontrolled cold water entering in response to the load requirements, and the *manipulated variable* is the amount of steam supplied to the tank. The *system lag* is the elapsed time from the time that heat is added to the tank to the time a response is detected by the water temperature sensor. The steam inlet control valve is a *continuous control* that may be set at any point in its range from full-off to full-on; it is not simply an ON/OFF control as might be found in a discrete control system.

In continuous process control, a setpoint provided by the operator is continually compared to the controlled variable output. A difference or error signal is generated which is proportional to the difference between the setpoint and the controlled variable output. This error signal is used by the controller to modify the output control module signal with an amplitude and polarity designed to reduce the amount of error.

A continuous process usually has the controlled variable compared to a setpoint. Any error between the two is amplified to generate an output to change the controlled variable and reduce the error.

**Figure 6-1.
Continuous Process
Control**

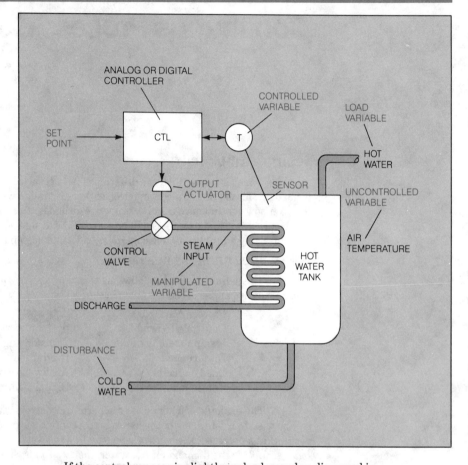

If the control process is slightly underdamped as discussed in Chapter 2, then the response to a change in setpoint will be as shown in *Figure 6-2a* and to a disturbance as shown in *Figure 6-2b*. That is, after the initial response to the change in a setpoint or to a disturbance, the process will stabilize at the new value with only a slight overshoot, but without sustained oscillation. If properly designed, the system will reduce the error to a level which ensures an acceptable product output.

CONTROLLER TYPES

As discussed in Chapter 2, there are four different types of controller systems commonly used for continuous process control. They are:

1. Proportional (P)
2. Proportional plus Integral (PI)
3. Proportional plus Derivative (PD)
4. Proportional plus Integral plus Derivative (PID)

The components of these will be discussed individually first, then the composite modes will be discussed. The discussions will be based on analog methods.

**Figure 6-2.
Closed-Loop Critically-
Damped Response for
Proportional System**

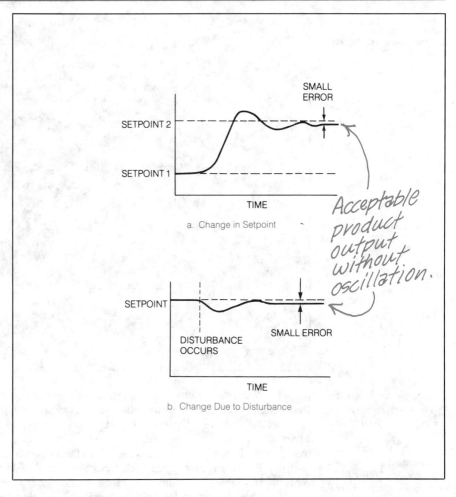

a. Change in Setpoint

b. Change Due to Disturbance

Acceptable product output without oscillation.

Proportional Control

In a proportional (P) control system, the output signal to change the controlled variable is proportional to the error signal.

The equation shown in *Figure 6-3a* is solved by the proportional controller and the response to an error is plotted in *Figure 6-3b*. If the output is greater than the setpoint, a negative error is input to the controller to cause the output to decrease and the error to get smaller. The opposite happens for an output less than setpoint. The equation is a series of straight lines with slope equal to the proportional gain factor, K_p, and with a bias along the Y axis of $H(0)$, the nominal setpoint. It is desirable to have the system designed so that the nominal setpoint is near midrange to allow a linear response to an error condition in either direction over a wide range. If the setpoint is near the end of the linear range, very little proportional action will be possible in one direction and for this correction it will act much like an ON/OFF controller. In operation, the output of the controller will be a bias amount (to move the setpoint to midrange) plus or minus an amount which is proportional to the difference between the setpoint and the control variable (the error).

**Figure 6-3.
Proportional Control
Output Response Versus
Error**

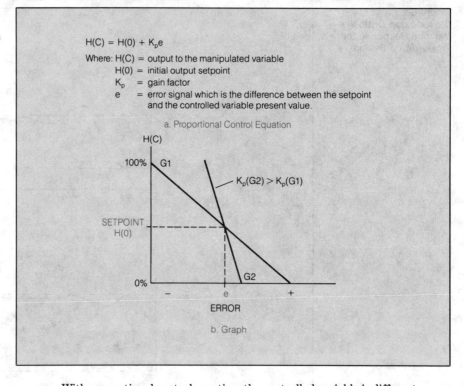

$H(C) = H(0) + K_p e$

Where: $H(C)$ = output to the manipulated variable
$H(0)$ = initial output setpoint
K_p = gain factor
e = error signal which is the difference between the setpoint and the controlled variable present value.

a. Proportional Control Equation

$K_p(G2) > K_p(G1)$

b. Graph

As a result, as the error correction gets closer to the set point the smaller the correcting signal becomes, and therefore, at equilibrium there will be a small constant error to provide a constant control signal.

With proportional control, anytime the controlled variable is different from the setpoint value, the controller will change the manipulated variable in the direction to correct the error. A problem with proportional control is that if a permanent disturbance occurs, this type of control will not be able to return the controlled variable to the setpoint value. The reason is that this type of control relies on an error signal to change the manipulated variable. As the controlled variable approaches the setpoint, however, the error signal will diminish which in turn will reduce the manipulated variable correction signal. The result is that the control variable will reach an equilibrium point as shown in *Figure 6-2* which will be somewhere below the setpoint and result in a constant error at equilibrium. If an attempt is made to eliminate the error by increasing the gain of the control loop, the control system will tend to go into sustained oscillation. The only way to eliminate this error using only the proportional control mode is to manually adjust the setpoint bias $H(0)$ to the new equilibrium point.

Analog Proportional Control

The operational amplifier circuit of *Figure 6-4* provides the proportional control of the equation of *Figure 6-3*. The equation is the sum of two linear quantities, therefore, a simple summing junction of an op amp can be used to solve the equation. One input has the setpoint value and the other input the error value. The gain factor K_p of the error input for the equation is R2/R1. The setpoint input gain is R2/R3, but since R3 = R2, the gain is 1 and the output equals the input. This, of course, is a solution in its simplest form and different gain factors probably would be used for a specific output controller or application. Typically, the gain factor can be adjusted by the operator if R2 is a variable resistor. Thus, the proportional gain can be set for each specific application when the system is installed.

**Figure 6-4.
Proportional Analog
Control Implementation**

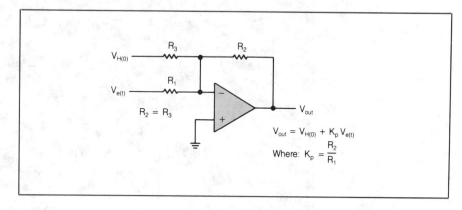

$$V_{out} = V_{H(0)} + K_p V_{e(t)}$$

$$\text{Where: } K_p = \frac{R_2}{R_1}$$

An operational amplifier with two inputs, one the setpoint and the other the error is a very useful circuit for analog proportional control systems.

One example of a proportional-only application would be a synchro positioning system as shown in *Figure 6-5a*. A synchro transmitter is a device that changes shaft position into electrical signals. A synchro receiver reverses the process. A synchro transmitter is attached to the output shaft of a system and sends phased ac signals to a remote indicator (synchro receiver) at a control station to indicate the shaft position. The output from the sending synchro can be thought of as a voltage that continually varies depending on the shaft angle. In actual practice, it is a sine wave and the voltage output is proportional to the difference in the phase angle between a reference sine wave and the generated sine wave. As a result of coupling the output of the sending synchro to the receiving synchro, the receiving synchro will follow the sending synchro in position. This is used often in extrusion forming and rolling mills in the steel industry. This is essentially a proportional control system which indicates shaft position. It is not very effective if large changes in velocity occur since it cannot respond well. This is analogous to the equilibrium offset error discussed above.

In other applications, limit switches that activate particular operations are activated by cams that are attached to the shaft of a remote receiving synchro as shown in *Figure 6-5b*. The particular operation desired is selected at the central station by setting the sending synchro.

Analog Integral Control

The control equation for the integral controller is given in *Figure 6-6a*. T_1 is an increment of time $(t_{i+1} - t_i)$ and e is the error over that increment $(e_{i+1} - e_i)$. The error signal is multiplied by the length of time the error is present. The errors for each small increment of time are summed together. In mathematics this is called integrating and is indicated by the $\int$ sign. The final integrated value has been continually summed and added to the previous signal until enough change is requested to bring the error to zero. At this point, the

**Figure 6-5.
Proportional Position
Sensor System**

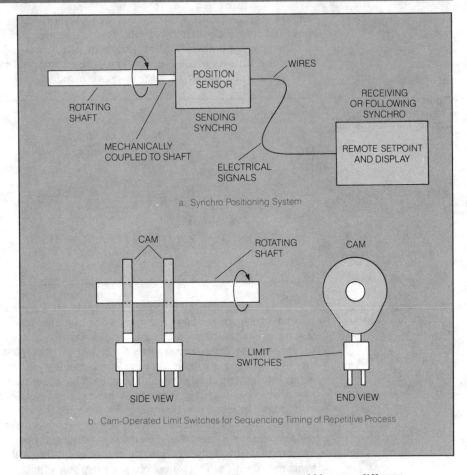

a. Synchro Positioning System

b. Cam-Operated Limit Switches for Sequencing Timing of Repetitive Process

Operational amplifiers also are very useful for providing the integrating function for integral (I) control in analog systems.

signal will cease to change, but the controller output could be very different from the initial setting, unlike the proportional control example, which always tries to return the nominal manipulated variable setting to the original setpoint as the error is reduced.

Operational amplifier circuits with the appropriate feedback components also can be used to provide integral control. An example circuit and its equation is shown in *Figure 6-6b*. This equation corresponds to the equation of *Figure 6-6a*. The setpoint V_{sp} is a constant independent of time so it is summed into a simple gain stage similar to $V_{H(0)}$ in the proportional controller example. The operation of the first op amp stage in this circuit initially delays the effect of the error signal on the controller output because all of the output change is fed back to the input through the capacitor C with a polarity to oppose the change. However, as the error persists, the negative feedback is reduced as the capacitor charges and the output is continually more affected by the error signal. This action ultimately will eliminate the error in the control variable assuming that it can be properly controlled by the equation being used.

**Figure 6-6.
Analog Integral Control
Implementation**

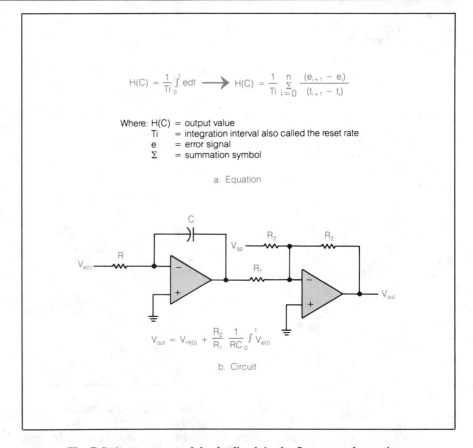

$$H(C) = \frac{1}{Ti} \int_0^t edt \longrightarrow H(C) = \frac{1}{Ti} \sum_{i=0}^{n} \frac{(e_{i+1} - e_i)}{(t_{i+1} - t_i)}$$

Where: H(C) = output value
Ti = integration interval also called the reset rate
e = error signal
Σ = summation symbol

a. Equation

$$V_{out} = V_{H(0)} + \frac{R_2}{R_1} \frac{1}{RC} \int_0^t V_{e(t)}$$

b. Circuit

The RC time constant of the feedback in the first stage determines the effect that the integral portion will have on the output. The longer the time constant, the less effect the integral will have. Conversely, as the time constant is shortened, the integral will have a greater effect. T_i is called the integration time, but it is often referred to as the reset rate by the control engineer. The reset rate is the reciprocal of the integration time; therefore, the longer the time constant, the lower the reset rate, and, again, the less effect the integral mode has on the outcome.

Analog Derivative Control

Another useful control mode is derivative control and it too can be provided by using operational amplifiers. *Figure 6-7a* is the control equation solved by the circuit shown in *Figure 6-7b*. The equation shown in *Figure 6-7b* is the same equation in terms of the circuit components, except that the control variable voltage, V_I, is substituted for the error signal, e, and RC is substituted for $1/T_d$. Notice the positions of R and C in this circuit are exchanged from those in *Figure 6-6*. This arrangement has the effect of reacting only to the changes in the control variable rate. As a result, the arrangement has a very poor response to very slow process parameter changes.

Derivative (D) controls monitor the rate at which the process control variable changes rather than the actual error, this provides quick, stable changes. Operational amplifier circuits are very useful in providing this function in analog systems.

**Figure 6-7.
Analog Derivative
Control Implementation**

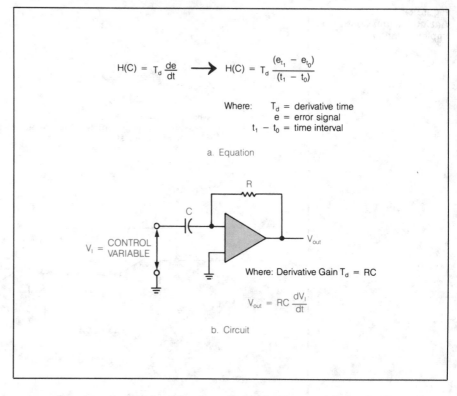

$$H(C) = T_d \frac{de}{dt} \longrightarrow H(C) = T_d \frac{(e_{t_1} - e_{t_0})}{(t_1 - t_0)}$$

Where: T_d = derivative time
e = error signal
$t_1 - t_0$ = time interval

a. Equation

$V_I = $ CONTROL VARIABLE

V_{out}

Where: Derivative Gain T_d = RC

$$V_{out} = RC \frac{dV_I}{dt}$$

b. Circuit

By properly choosing the values, the derivative control circuit can anticipate the system errors by reacting quickly to changes in the rate of change of a variable. This has the effect of making the control response faster and maintaining the setpoint better. In actual practice, as shown in *Figure 6-7b*, the control variable is used rather than the error signal itself in order to avoid incorrect response to impulses.

In this mode, the gain is set by adjusting the time constant of the RC network, typically by adjusting the resistor value. The larger the resistor value, the longer the time constant, and the greater the effect that the derivative circuit will have on the control response. Conversely, the shorter the time constant, the less effect on the response. Like the integral mode, this time constant is commonly referred to by its reciprocal which is called the rate control. The higher the rate control, the less effect the derivative will have on the control response.

COMPOSITE MODES

The combining of P, I and D control modes for specific applications provides more accurate and stable systems than any of the systems separately.

When each of the three continuous control modes is used separately, each has problems which limit its effectiveness. In most cases, one of the problems typically can be controlled by one of the other modes. Thus, it might be reasoned that using two or more of the modes in combination would provide better control. In fact, this is done, but the system must be designed and adjusted (tuned) so that each mode is used when it is most effective.

Usually, for a properly tuned system, the derivative control dominates in the short term and the proportional control is most effective during the short to medium time period following a step response. The integral control helps in the long term to null out all sensed errors and to reset the setpoint as uncontrolled parameters change.

PI Mode

Proportional-integral (PI) control is used in situations requiring significant changes in the set-point due to large amounts of change in the load.

The proportional-integral (PI) control probably is the most widely used composite control in industry. The PI control is useful in situations where there are large load changes which will require significant changes in the setpoint. The integral portion can handle this very well and also eliminates the offset error of the proportional-only control. The use of the proportional mode results in a faster response (than integral alone) to load changes. The PI control is most effective in systems where the load changes occur slowly because fast load changes can cause system instability if the integration time is not set properly. Another advantage of the PI control is that the integral portion eliminates the offset error of the proportional-only control. Thus, the proportional band can be set wider (lower proportional gain factor) which will help in lowering the response overshoot of the system. The equation satisfied by the PI mode is a combination of the proportional control and integral control equations. It is:

$$H(C) = H(0) + K_p e + K_p \frac{1}{T_i} \int_0^t edt$$

Where: e $\quad$ = error signal
$\quad\quad\quad$ K_p $\quad$ = proportional gain
$\quad\quad\quad$ T_i $\quad$ = Integral time
$\quad\quad\quad$ H(0) = Setpoint

As shown by this equation, the total integral gain is affected by the proportional gain factor, K_p, but the integral gain factor, T_i, can be independently controlled to obtain the desired total integral gain.

PD Mode

Proportional-derivative (PD) controls are less common but are useful when small, rapid changes in the load are likely to occur.

The proportional-derivative (PD) control is not used widely in industrial control applications since it does not eliminate the offset error of the proportional control mode. It responds only to second order changes, that is, changes to the error signal, not the error signal itself. The PD control can be useful in systems that have rapid changes in load since PD control tends to anticipate the total error by responding to how fast the error signal is changing. Typically, the faster the rate of change of the error, the greater the total error will be.

PD control is used in motor servo sytems and in systems that have small, but quick, process parameter changes. The equation for the PD control is:

$$H(C) = H(0) + K_p e + K_p T_d \frac{de}{dt}$$

Where: e $\quad$ = Error signal
K_p $\quad$ = Proportional gain
T_d $\quad$ = Derivative time
$H(0)$ = Setpoint

Note that interaction between the proportional response and the derivative response is indicated by the multiplication of K_p (the proportional gain factor) and T_d (the derivative factor). The interaction can be controlled by appropriately setting the two gain factors since they are set independently.

PID Mode

Overall, the most flexible process control is the proportional-integral-derivative (PID) controller. While more difficult to set up, PID controllers can be used in almost every operation requiring control.

The composite control which can meet virtually every industrial process control need is the Proportional-Integral-Derivative (PID) Control. It was first developed using pneumatic logic. The control has gained wide acceptance and is the common measurement benchmark for comparison of all continuous process control loops. The PID mode can be used in virtually every application by properly tuning each loop to the specific time constants and gain factors required.

As one might assume, the amount of interaction in setting up the parameters in a PID control system is greater than for any other control discussed so far. The equation for the PID control is:

$$H(C) = H(0) + K_p e + K_p \frac{1}{T_i} \int_0^t edt - K_p T_d \frac{dcv}{dt}$$

Where: K_p $\quad$ = Proportional gain
T_i $\quad$ = Integral time interval
T_d $\quad$ = Derivative time interval
e $\quad$ = Error
cv $\quad$ = Control variable
$H(0)$ = Setpoint

As shown by the equation, the amount of adjustment that is available and the amount of interaction for each adjustment requires that the sytem be understood well in order to get the optimum response of the control loop; it is not sufficient to just use the PID mode, it must be set up properly.

Naturally, there is an optimum point for each of the gain factors in the system; too small an effect means poor system control, too large an effect can result in oscillations or even an unstable system where the amplitude of the oscillation increases over time.

Because extra effort is required to set up the system properly, it should be used only if the extra effort will produce an improvement in the system control response and this improvement is needed. If it is not needed, the PI mode may be an easier mode to use.

In most dedicated continuous process controllers, the PID equation is already solved and the control engineer need only determine the system parameters to enter into the appropriate loop equations as specified by the controller. If the full PID controller is not needed, null values may be put in for any of the system parameters and that portion of the PID composite eliminated.

DIGITAL METHODS

All of the control capabilities provided by linear, proportional control circuits can be performed by digital systems.

Instead of using analog circuits to solve the process control linear equations, they can be solved algebraically using a digital computer. The analog signals are converted to binary data and operations like addition, subtraction, multiplication and division can be used to solve any of the process control equations that have been considered so far. Software algorithms must be developed to handle operations like derivatives, integrals, square root, etc. Once this is done, the algorithms can be used as often as necessary in order to solve the process equations.

A graphical example of computing the rate of change of a control variable used for the derivative control loop is shown in *Figure 6-8*. The method used in an actual controller is to divide the time into discrete amounts and look at the error at each point. The derivative is simply the slope of each one of these line segments and can be computed for each line segment by taking the control variable difference and dividing by the time interval.

**Figure 6-8.
Graphical Solution of
Derivative**

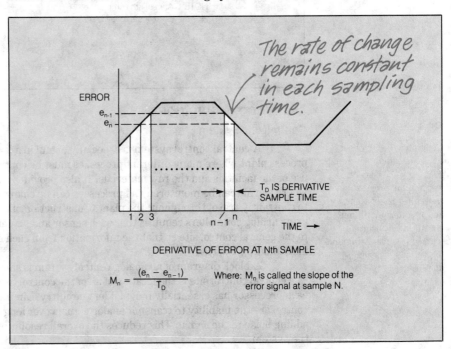

The rate of change remains constant in each sampling time.

ERROR

e_{n-1}
e_n

T_D IS DERIVATIVE SAMPLE TIME

1 2 3   n−1 n

TIME →

DERIVATIVE OF ERROR AT Nth SAMPLE

$$M_n = \frac{(e_n - e_{n-1})}{T_D}$$

Where: M_n is called the slope of the error signal at sample N.

Typically, integration is handled in a similar fashion in an actual controller. A graphical example is shown in *Figure 6-9*. The time is divided into fixed intervals and the error is measured at each interval. This error is then assumed to have been constant during that time interval. Thus, integration amounts to adding up the areas in a series of rectangles. This summation is called the time integral of the error, hence, the name of integral control.

**Figure 6-9.
Graphical Solution of
Integration**

Integral and derivative control functions are provided in digit control systems by dividing the error response into small time intervals. Adding areas in the intervals provides integration; calculating rate of change of error provides the derivative.

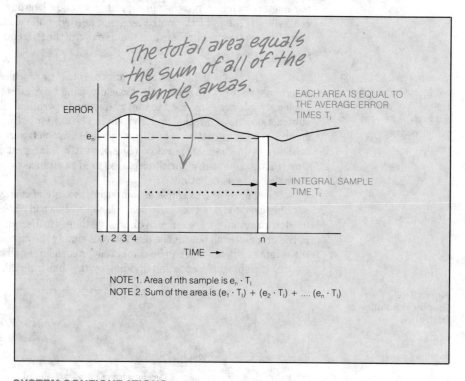

The total area equals the sum of all of the sample areas.

EACH AREA IS EQUAL TO
THE AVERAGE ERROR
TIMES T_I

ERROR

e_n

INTEGRAL SAMPLE
TIME T_I

1 2 3 4   n

TIME →

NOTE 1. Area of nth sample is $e_n \cdot T_I$
NOTE 2. Sum of the area is $(e_1 \cdot T_I) + (e_2 \cdot T_I) + \dots (e_n \cdot T_I)$

SYSTEM CONFIGURATIONS

Centralized System

Centralized control systems have all error signals sent back to the central location for processing; therefore, their accuracy and resolution are limited because of difficulties found in transmitting analog signals over long distance.

A central control system is especially useful in a large interdependent process plant where many different processes must be controlled for efficient use of the facilities and the raw materials. It also provides a central point where alarms can be monitored and processes can be changed without the need to constantly travel throughout the plant. Construction of this type system using analog controllers requires that each sensor and actuator be wired back to the central control site so that the information from each remote site can be individually displayed.

A block diagram of an analog central system is shown in *Figure 6-10*. The central control site is the location of all of the control processors although each processor has essentially independent circuitry. Limitations of this type of control are the inability to transmit analog signals over long distances without adding noise to the signal. This reduces the overall resolution and accuracy of the system.

**Figure 6-10.
Analog Central System
Block Diagram**

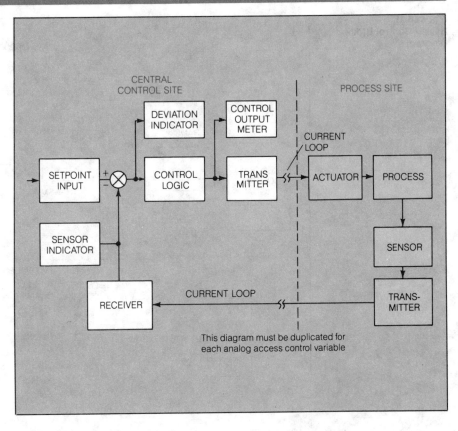

This diagram must be duplicated for
each analog access control variable

Hybrid Control

Hybrid control combines analog control advantages with digital communication control features.

Hybrid control is a mixture of analog control loop processing and digital signal transmission to take advantage of the ease of use of the analog control and the advantage of noise immunity and reduced wiring expense of digital serial data transmission. In this configuration, local process control is handled by the analog control elements, and a digital computer-based central control system provides the ability to enter setpoints into each controller from one location. Each controller is interfaced through a digital transmission link. A block diagram of this type of system is shown in *Figure 6-11*.

This system allows the computation of complex calculations or the solving of complex algorithms. The results are changes in the setpoints for different local processors. This is especially useful when significant process interaction is present and the relationship between the variables is not straightforward.

**Figure 6-11.
Hybrid Control Block
Diagram**

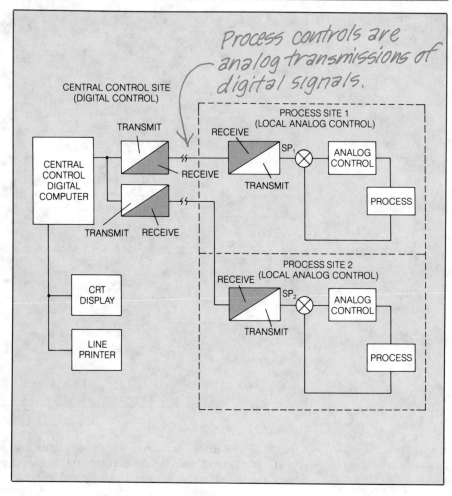

Process controls are analog transmissions of digital signals.

Distributed Digital Control

In distributed digital control facilities, each local system control point has its own digital processing unit, monitored by a central computer. If trouble occurs in the central computer, the local system still operates.

Distributed digital control permits the distribution of the processing for each task among several control elements. Instead of just one computer located at a central control point doing all the processing, each local loop controller has digital processing capability. A block diagram of a distributed control system which allows both local and remote monitoring points is shown in *Figure 6-12*. This system also allows changing of setpoints and cascading of different local control loops by passing the information through the central control computer. All of the local loop information can be passed to the central control computer so that it can be monitored easily by the operator. At the same time, each individual control loop operates independently from the others so that any communications failure or central computer overload does not affect the operation of any of the real-time process control loops. The only effect would be the inability to monitor the loops, or to pass information from one loop to another, or to change the setpoints from the central control site.

**Figure 6-12.
Distributed Digital
Control System Block
Diagram**

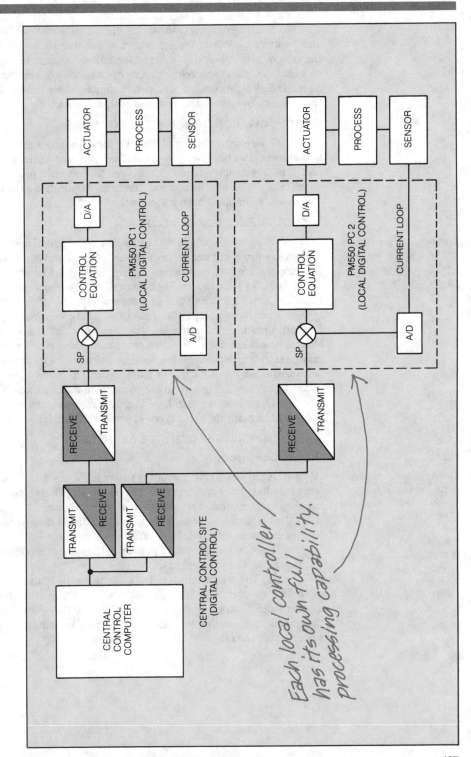

*Each local controller
has its own full
processing capability.*

Each digital processor solves the control equations for its local loop under software control. Analog sensors and analog actuators are interfaced to the system with A/D and D/A converters. Many times other logic functions can be added easily and new requirements can be added by reprogramming one or all of the digital processors. As a result, digitally-based controllers represent the present direction in the design of continuous control.

CONTINUOUS CONTROL SYSTEM EXAMPLES

There are a number of methods for designing a continuous control loop system. Two types of systems that can perform continuous process control are: (1) microcomputer-based systems and (2) dedicated process controllers which take the place of analog counterparts. Application examples of each type are presented in the following sections.

MICROCOMPUTER-BASED CONTROLLER

Commercially available microcomputers can be used to provide a continuous control system that, once programmed, permits many adaptations to different applications.

Here is an example of the use of an off-the-shelf microcomputer system to solve continuous control problems. This system example regulates the temperature for a manufacturing facility. A Texas Instruments TM990 microcomputer-based controller maintains the desired air temperature year-round for a manufacturing area and several adjacent offices. One disadvantage of using a microcomputer system like this is that the software (the continuous control program) must be developed by the user without as many programming shortcuts as there might be with a programmable controller, but soon there may be standard process control solution software for a microcomputer and the software development job will be much easier. Of course, once the software is developed, modifying it to adjust to many new applications is much easier. The use of a well-chosen high-level language for this software would permit it to be used on many different types of processors with only minor adaptations.

AREA TEMPERATURE CONTROL EXAMPLE

System Description

The continuous area temperature control system monitors the input air sensor and the output air sensor as well as regulating the flow in the hot and cold water coils to maintain a constant temperature of the air blowing over the coils.

A physical system layout is shown in *Figure 6-13*. There is one large manufacturing area and six adjacent offices. The temperature of the air supplied to the manufacturing area must be maintained at 68° ±1°F (50.2° ±1.8°C) at all times while each office temperature is maintained at or above 68°F as determined by individual office thermostats.

The continuous control portion of the system is responsible for holding the air temperature at the air duct output into the manufacturing area at 68° ±1°F. The sensor inputs to the system are the temperature inputs of the input duct air after the heater, T_H, and output duct air after the cooler, T_C. The system control outputs operate control valves for the hot-water heater coil and the chilled-water cooling coil. All of these inputs and outputs are continuously variable over the range of interest.

**Figure 6-13.
Temperature Control
Layout**

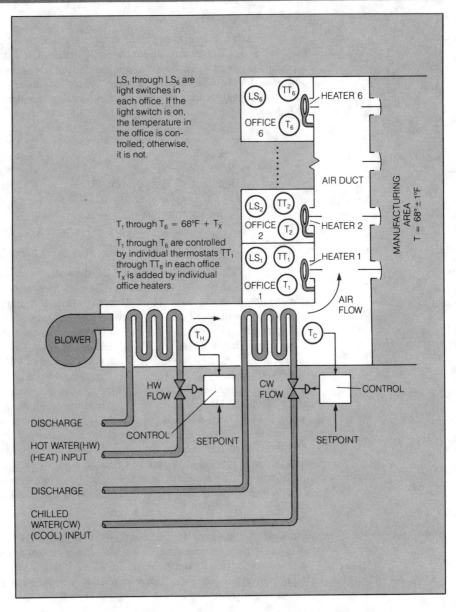

LS$_1$ through LS$_6$ are light switches in each office. If the light switch is on, the temperature in the office is controlled; otherwise, it is not.

T$_1$ through T$_6$ = 68°F + T$_x$

T$_1$ through T$_6$ are controlled by individual thermostats TT$_1$ through TT$_6$ in each office. T$_x$ is added by individual office heaters.

HEATER 6

OFFICE 6

AIR DUCT

HEATER 2

HEATER 1

AIR FLOW

MANUFACTURING AREA
T = 68° ± 1°F

BLOWER

HW FLOW

CONTROL

CW FLOW

CONTROL

SETPOINT

SETPOINT

DISCHARGE

HOT WATER(HW)
(HEAT) INPUT

DISCHARGE

CHILLED
WATER(CW)
(COOL) INPUT

        The individual offices are controlled using the limit-cycle control method. Electrical resistance heaters are located at the duct entrance to each office so that individual offices can be maintained at a temperature higher than the 68°F of the air duct. Each office light is monitored to determine if the office is occupied. If the light is on, the thermostat setting is maintained; otherwise, the heater is turned off to conserve energy.

Controller Hardware

Each of the sensor outputs for the temperatures that are controlled (T_1 through T_6, T_H and T_C) and for the light switches, is a specific computer input and is assigned a unique I/O address for the use of the controller. The addresses for the flow valve and heater control outputs are assigned in the same way. The total I/O can be divided into analog and digital signals as follows:

Inputs	Outputs
14 Analog	2 Analog
Room Temperatures (6)	Hot Water Flow (1)
Heating Coil Temperature (1)	Chilled Water Flow (1)
Cooling Coil Temperature (1)	
Room Thermostats (6)	
6 Digital	6 Digital
Light Switches (6)	Room Heaters (6)

The TM990 product line has standard plug-in preassembled circuit boards which can be used to meet these I/O requirements. Using such preassembled boards allows the system to be up and running very quickly because all the detailed system hardware design is already done. The system configuration in *Figure 6-14* shows that three boards are required for this application. They are the TM990-101 CPU board which has a BASIC interpreter in onboard ROM, two serial ports and up to 16K of memory; a TM990-305 memory and I/O expander for additional memory and a cable connection to receive inputs and activate outputs from and to 5MT modules; and a TM990-1241R analog board which has 16 analog inputs and two analog outputs. This board contains the A/D converters for the analog input conversions and the D/A converters for the digital output conversions. Additional hardware requirements are the 5MT industrial I/O modules, a card cage, a power supply, and possibly a CRT. The 5MT modules accept ac voltage inputs and provide standard dc logic levels to the computer or receive standard dc logic levels from the computer and provide industrial ac power (115 to 230 Vac) control.

The software for the system can be developed on an external system or directly on this system by adding the following: a CRT, monitor control software to help with accepting programming commands, a means of storing the program during its development, and an EPROM programmer capability to put the system control software into ROM (firmware) so it can be installed in the microcomputer.

**Figure 6-14.
Control System
Configuration**

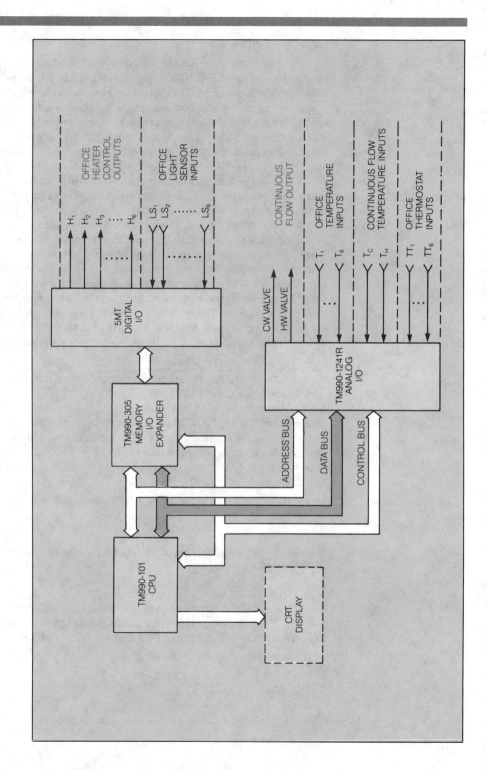

Controller Software

The software to run this control system is written completely by the user in the high-level BASIC language by using POWER BASIC® available with the TM990 products. The system program is divided into two separate operations for this application; one for the continuous process loop for controlling the manufacturing area temperature, and one for the limit-cycle control of the individual office temperature. The primary purpose is to maintain the duct air at the prescribed temperature for the manufacturing area, and then condition the air further for the offices. If the temperature in the office is below the thermostat setting, heat will be added until the thermostat setting is reached. The heater will cycle on again when the temperature is below the thermostat setting.

The software is set up so that it scans the appropriate sensors at a rate which is much faster than the output or input can change significantly. The office sensors are scanned every 10 seconds and each scan requires CPU time of less than 500 milliseconds so this does not take a significant amount of computer time. The continuous process sensors for the duct air temperature control also are scanned at a predefined rate. This rate, however, will affect the operation of the loop. The continuous process loop is the primary loop and it takes priority over the secondary office loop, but this will not be a significant problem in this application since the microcomputer is idle much of the time.

Input/Output Conversion Software

The temperature sensors provide an input signal of 4 to 20 mA over their linear range between 32° and 96°F (0° and 78.2°C), where 32°F causes 4 mA to flow in the loop and 96°F causes 20 mA to flow. A scaling resistor is placed on the TM990-1241R board A/D input to convert the current flow to a voltage within the operating range of the input. Since these board inputs are designed for operation over the range of 0 to 10V, a 500 ohm resistor was chosen to convert the 4 mA signal to 2 volts and the 20 mA signal to 10 volts. The input scale factor (SF) is then:

$$
\begin{aligned}
SF &= (96 - 32°F)/(10 - 2V) \\
&= 64/8 \\
&= 8°F/V
\end{aligned}
$$

The scaling and normalization of the input V1 can be handled with this one line of POWER BASIC code:

$$\text{LET T1} = (V1 - 2)*8 + 32$$

Each of the other inputs can be scaled in a similar manner. For instance, if the individual office thermostat linear range is chosen to be between 70° and 86°F (52.2° and 68.2°C), the scaling and normalization is:

$$
\begin{aligned}
SF &= (86 - 70)/(10 - 2) \\
&= 16/8 \\
&= 2°F/V
\end{aligned}
$$

®POWER BASIC is a trademark of Texas Instruments Incorporated

and the POWER BASIC statement for input V11 is:

$$\text{LET } S1 \ = \ (V11-2)*2+70$$

The analog outputs consist of two current loops which are provided on the 1241 boards. These outputs drive servo type valve actuators which operate from fully open at 4 mA to fully closed at 20 mA. In the 1241 board design, this corresponds to a number between 0 and 4095. In other words, the computer outputs the binary code of a number between 0 and 4095 representing a valve control setting from fully open to fully closed. The D/A converter on the 1241 board converts the number to a current between 4mA and 20mA to operate the valve.

Office Control Software

The limit-cycle software uses a pure digital system to control each of the offices individually. The present temperature is compared to an average of the present temperature and three previous temperatures and heat is added if the temperature is too low.

The digital inputs and outputs are used to control the office heating elements. If the inputs sense the presence of a voltage across the lights, it is assumed the office is occupied. The direct ac inputs to the computer boards are limited to 30 Vac, therefore, a 5MT input module capable of handling 115 Vac is used to sense the voltage across the light. The system output then controls the corresponding office heater element. The outputs are connected through a 5MT output module which supplies electrical isolation through optical coupling and is capable of switching 115 Vac to turn the heater on or off.

The office control software flow diagram is shown in *Figure 6-15*. The operation of the office control portion of the software is to monitor the temperature setting and compare it to the average value of the particular office temperature. The previous three readings and the present reading are averaged to reduce the effect of any noise in the reading and prevents unnecessary ON/OFF toggling of the heater power controls. This average temperature is then compared to the setpoint $\pm 1.5°F$ (which gives a 3°F deadband) to determine if the heater should be on or off. If the office temperature is lower than the set point by 1.5°F, the heater is turned on for that office (or if already on, left on). If the heater is on and the office temperature is higher than the set point by 1.5°F, the heater is turned off. This cycle is repeated for each office until all offices have been scanned. When all six offices are scanned the program starts over.

Additional software can easily be incorporated to limit the maximum temperature setting or provide real time clock control of the hours that the heaters are allowed to run rather than by sensing occupancy. These are simple background tasks which will not significantly load down the computer or prevent it from performing its primary function of controlling the duct air temperature.

**Figure 6-15.
Office Temperature
Control Software Flow
Diagram**

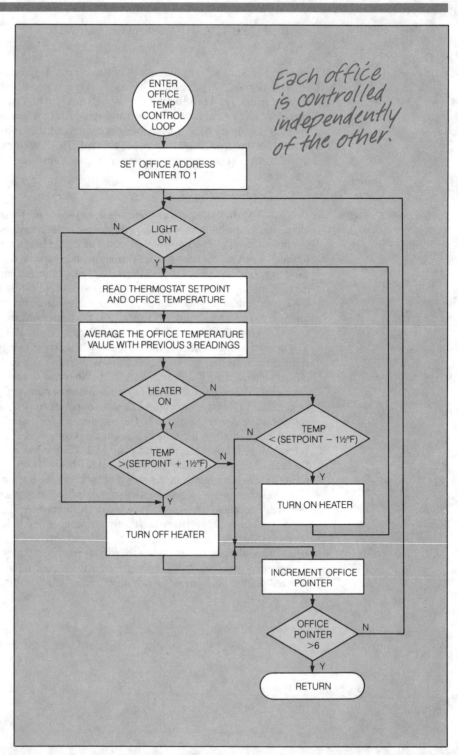

Manufacturing Area Control Software

The software controls the temperature in the manufacturing area by first adding heat to bring the temperature to above the set temperature for a coarse control and then add cooling to fine-tune the temperature to the set value.

The operation of the temperature control process for the duct air temperature is such that both heating and cooling always will be required in normal operation. The procedure is to first heat the air to slightly over the desired temperature with the heating coil, then cool the air to the desired temperature. This will always result in some cooling requirement and will maintain the cooling valve in its linear region to provide better control.

A software flow diagram of this control loop is shown in *Figure 6-16*. The software operation of the loop itself is very simple: It senses the input air temperature and adds heat if it is below 69°F; it senses the output duct temperature and adds cooling if it is above 68°F. The resultant control, however, is relatively sophisticated. The equation that is used to compute the output value for the control to the valve is the equation for a proportional integral control. It is:

$$CTL = H(0) + K_p e + K_p \frac{1}{T_i} \int_0^t e\,dt$$

where: e = the error
K_p = proportional gain constant
T_i = integral time (reset rate).

K_p is modified by entering into the system the initial parameters based on the physical parameters of the system. The integral gain $K_p(1/T_i)$ is a function of K_p and the rate at which the integral portion of the equation is updated. The shorter the time, the greater the effect of the integral portion.

In the equation:

$$CTL = K_p * e + (I - [K_p/T_i] * e)$$

shown in *Figure 6-16*, the actual integration over the integration time T_i is $(K_p/T_i) * e$. This then affects the last value of I by summing in the $(K_p/T_i) * e$ value. The new value of I is then updated in its storage location so it can be used for I in the next integration time T_i.

The total integration value I is held constant in the control equation when the controlled variable CTL is either 0 or the full range of 100% (1 in this case). These end points are called "in saturation". When CTL is at 0, the I term is clamped at 0 until CTL comes out of saturation back into its linear range. Similarly, when CTL is 1, I is clamped at 1 until CTL comes out of saturation back into its linear range. If the errors were simply summed together forever, some conditions (like at startup) might cause the algorithm to stay in saturation until long after the setpoint had been reached. This would require a much longer time for the loop to settle.

**Figure 6-16.
Continuous Control
Software Control Loop**

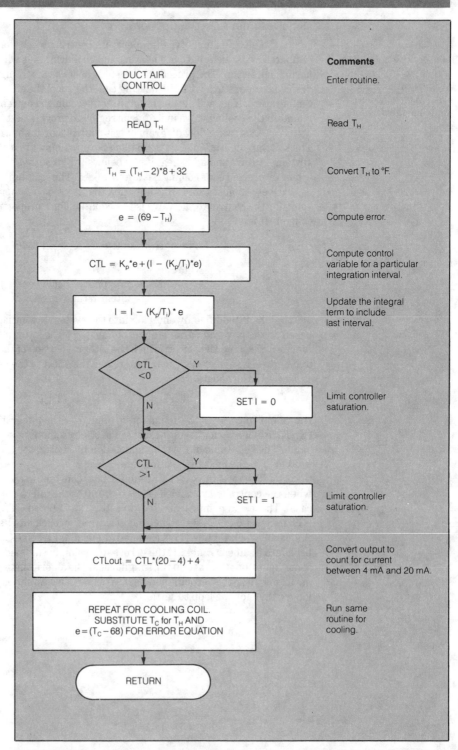

Flowchart	Comments
DUCT AIR CONTROL	Enter routine.
READ T_H	Read T_H
$T_H = (T_H - 2)*8 + 32$	Convert T_H to °F.
$e = (69 - T_H)$	Compute error.
$CTL = K_p*e + (I - (K_p/T_i)*e)$	Compute control variable for a particular integration interval.
$I = I - (K_p/T_i)*e$	Update the integral term to include last interval.
CTL <0 — Y → SET I = 0	Limit controller saturation.
CTL >1 — Y → SET I = 1	Limit controller saturation.
$CTLout = CTL*(20 - 4) + 4$	Convert output to count for current between 4 mA and 20 mA.
REPEAT FOR COOLING COIL. SUBSTITUTE T_C for T_H AND $e = (T_C - 68)$ FOR ERROR EQUATION	Run same routine for cooling.
RETURN	

After the heating control variable is set, the software continues in the same type sequence to set the cooling control variable. Each of the valves has a loop around it; the hot water valve loop acts as the coarse control and the cold water valve loop acts as the fine control. The cold water valve control, therefore, can have a higher gain than the hot water valve control. The sample rate for each loop is programmable as is the proportional gain so that both the proportional and integral responses can be optimized during operation.

Maximum control flexibility is available because the software allows changes in sample rate and proportional gain to be made.

The same procedure could be used to add the derivative response; however, in this application, the derivative response did not significantly affect the system performance and the amount of time required to determine the gain coefficients could not be justified.

The POWER BASIC statements (and their explanations) to implement the loop are shown in *Figure 6-17*. These are incorporated into a larger control loop which maintains the entire system software sequence of operation. In reality, the time delay rates would be written as a variable with another portion of software giving the control engineer the ability to modify the time delay rates from a keyboard or by some other method.

Figure 6-17.
BASIC Program for the Flow Diagram of Figure 6-16

		Comments
10	TH = INP(100)	Get the hot coil air temp and store in T_H.
20	TH = (TH − 2)*8 + 32	Scale T_H value into °F instead of a voltage.
30	E = 69 − TH	Compute the error in °F.
40	CT = KP*E + (I − (KP/TI)*E)	Solve the new control point value.
50	I = I − (KP/TI)*E	
60	IF CT < 0 THEN I = 0 ELSE GOTO 70	Prevent controller windup.
70	IF CT > 1 THEN I = 1 ELSE GOTO 80	Prevent controller windup.
80	CT = CT*(20 − 4) + 4	Scale the output value to correspond to the 4-20mA loop current.
90	OUT 120, CT	Output the computed value setting to the heat control valve.

Use a similar routine for the cooling coil, except substitute TC for TH and E = TC − 68 for error in Step 30. Of course, the gain constants may differ in actual value for the two loops and would be given different names.

DEDICATED PROCESS CONTROLLER

One example of a dedicated process controller is the PM550 programmable controller manufactured by Texas Instruments. Even continuous processes usually require some discrete controls and this controller can perform both continuous process control and relay ladder logic control simultaneously. Using an analog control system, two separate controllers would have been required.

A TM5509 programmable
controller dedicated pro-
cess control system is very
versatile. A coffee bean
roating system is used as
an example.

Figure 6-18 shows PM550 system hardware and *Figure 6-19* shows a typical system block diagram. It consists of a central control unit, power supply, digital (discrete) and analog input/output modules, a timer/counter, a loop access module for monitoring, an I/O simulator, and a programmer. The programmer is used only to initially program the system or modify an existing program once the system is in operation. The unit can accommodate up to 512 discrete input or output points, up to 8 timer/counter functions and 8 PID loops. I/O units may be added as required up to the maximums given. The loop access module allows an operator to ask the system to display any of the appropriate parameters required and allows modification of these parameters.

The PM550 is used in a large number of industrial control environments from steel mills, to chemical process plants, to food processing applications. One application is dry roasting coffee beans as discussed in Chapter 2. Although this process is very old, a number of factors in recent times have increased the need for better control. The PM550 is ideal for providing the improved control.

**Figure 6-18.
PM550 Programmable
Control System**

**Figure 6-19.
Typical PM550 System
Block Diagram**

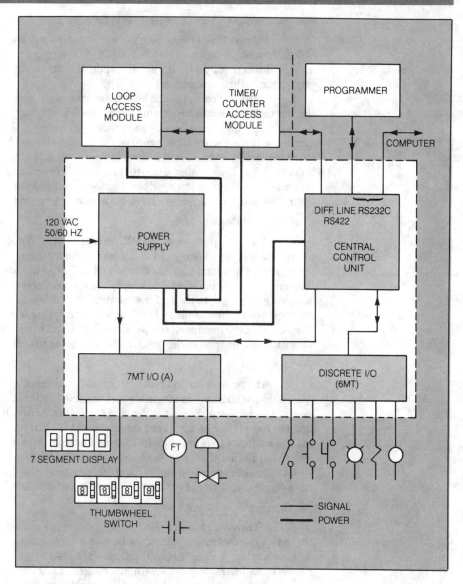

COFFEE ROASTING EXAMPLE

Maximum Energy Efficiency

Most commercial coffee processing is done using gas-fired heating systems. Up until 1973, high quality gas with a consistent content of BTU per cubic foot was available at reasonable prices. Today, the quality of the gas available is not as consistent and the cost has risen tremendously; therefore, the need to get maximum heat efficiency from the gas is now much more important. This has resulted in the need to obtain better control of the gas/air ratio in the combustion process.

Reduced Process Time and Raw Material Waste

Process time affects both the amount of energy used and the number of facilities required to process the product. The longer the process time, the more facilities required for the same output. If better control can reduce the process time, then production cost will be lower.

Today the cost of coffee beans is considerably higher than in the past; therefore, any errors in processing which result in spoilage or waste are more costly. In this application, waste and process time are closely related. The higher the temperature of the process, the shorter the process time and the lower the energy cost; however, if the temperature is too high, the coffee bean is destroyed and raw material waste is increased. In order to reduce process time while maintaining low spoilage, the temperature must be closely controlled to tight tolerances so that both process time and waste is minimized.

Fuel Demand Control Loop

The basic purpose of the PM550 controller is to maintain the temperature of the roasting bin at the correct temperature. This is accomplished by controlling the flow of fuel and air to the burner. The temperature of the bin is used to supply the overall setpoint to the system by supplying a setpoint to the fuel flow control loop. This loop must maintain the temperature as close as possible to the optimum drying temperature while ensuring that the maximum temperature limit is not exceeded. *Figure 6-20* shows a block diagram of the control system needed to control this process.

Fuel/Air Mixture

The process control must incorporate additional control loops to ensure that maximum gas efficiency is maintained and that minimum raw material losses occur. When the demand for heat increases, it is essential that the fuel and air be increased simultaneously to maintain the air/fuel ratio within safe limits to prevent inefficiency, flame extinction, or in the extreme case, an explosion. These possibilities are realities because of differences in the system dynamics between air flow and fuel flow controls. As a result, fuel/air cross-limiting is necessary to ensure some excess air at all times.

Air and Fuel Flow Control

Figure 6-21 is a simplified control diagram showing the sensor inputs at the top, the control outputs at the bottom and the control loops interacting with each other. A significant advantage of the PM550 system is that it can handle these interactive loops with just programming inputs.

The fuel flow set point for the PI controller at C of *Figure 6-21* results from the smaller of the signals 1 or 2 at A. 1 is the demand from the bin temperature input and 2 is the cross limiting input so that the fuel is limited by the air that's available.

**Figure 6-20.
Coffee Roaster Control
Diagram**

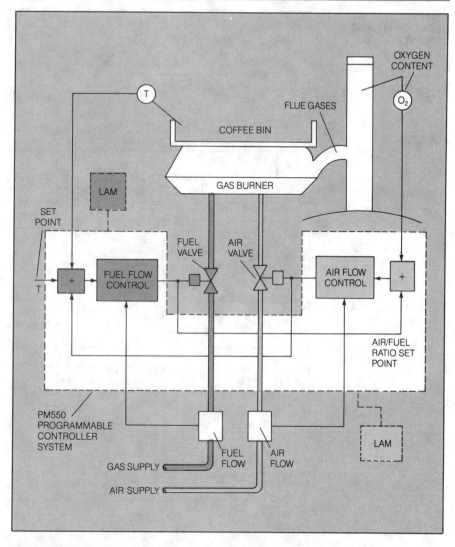

The air flow set point for the PI controller at D results from the larger of the signals 1 or 3 at B. 1 again is the demand from the bin temperature input and 3 is the cross limiting input so that the amount of air is maintained for the amount of fuel that's available or present. This arrangement ensures that the fuel flow decreases before the air flow and that the air flow increases before the fuel flow. The loops then compare the setpoints with the actual flow rates and, using the PI algorithm in the controller, solve for the correct actuator output.

Air Flow/Oxygen Correction

In addition, since the gas quality is not constant, a simple flow ratio is not sufficient to ensure maximum burn efficiency. The result is that the oxygen content of the exhaust gas in the flue stack is monitored to provide an input on

**Figure 6-21.
Combustion Control with
Oxygen Correction**

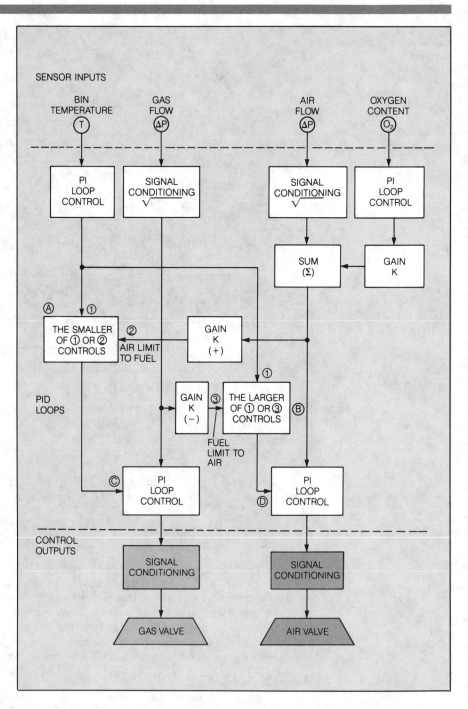

the efficiency of the combustion process so the control system can adjust for variations in the heat content of the fuel. This loop provides a setpoint input to the air flow control loop which adjusts the relative flow of air to fuel.

Loop Preparation

The programmer need only provide the control parameters to a programmable controller system and it will set up the PI loops as well as indicate the information required and the order this information is needed.

Now the real advantage of a programmable controller system like the PM550 can be demonstrated. The same flow diagram structure as shown in *Figure 6-16* for a PI loop must be set up for each one of the PI loops shown in *Figure 6-21*.

If the microcomputer-based system is used, a subprogram for each of the loops will have to be written; however, in the PM550, the PID loops are already established. The only requirement is that the gain parameters, the integral times (reset rate) and the differential times (rate control) be programmed into the system along with the addresses of where constants and variables are to be found, where outputs are to be located, the type calculations that are to be made, and the units or scaling factors that are to be used by the system.

Not only does the system set up the loops from the information, but it also prompts the programmer with questions so that the correct information is entered in the right sequence. This allows the control engineer to concentrate on the process and not on the programming. *Table 6-1* is a loop specification sheet. It provides a documented copy of the important system parameters and helps simplify the entry of the data by the programmer. The sheet is divided into four areas: Loop table memory, process variable specifications, control calculation specifications and alarm limits. The PM550 prompts the programmer in this same order. Here are the details of the areas.

Memory Allocation

Each loop requires both variable and constant data. Entries in the table provide pointers so that the PM550 knows where in memory to look for data when it is needed. This data storage is used for gain factors, set points, and other process variable data.

Process Variable

The PM550 will prompt for the address of the process variable input, whether there is an offset, and if a square root is required along with other questions relating to units and scale factors as listed on the specification sheet.

Control Calculations

Control loop tuning constants (proportional gain, reset time, and rate time (rate control) along with remote set point, output addresses, and polarity of gain are entered. The polarity of gain is determined by the answer to the question, "Reverse Acting?"; *Yes* means negative polarity. Various other options are also possible, but are beyond the scope of this book.

**Table 6-1.
Loop Specification Sheet**

Loop Description: _____ Gas Flow _____ PM550 Loop Number _2_____

MEMORY ALLOCATION

Tuning Constants in _V_ (C or V)

	Beginning Address	Table Length Tune in C	Table Length Tune in V	Ending Address
Constant Table	C200			C214
Variable Table	V200			V217

Are loop flags for alarms and mode switching allocated in the image register? _Yes_____

If yes, give beginning address: _____ CR201 _____
(10 successive locations will be used)

PROCESS VARIABLE   Address: ___V57 (A103)___

20% Offset: ___Yes___   Square root? ___Yes (V202)___

Special calculation? ___No___   If yes, give address: ___—___

Low Range = ___0___   High Range = ___1000 SCFH___

Engr. Units ___SCFH___   Transmitter ___FT-102___

CONTROL CALCULATIONS   Sample Time = ___0.5 sec.___

Remote Set Point? ___Yes___   If yes, give address: ___V218___

Special Calculation? ___No___   If yes, give address: ___—___

Lock — Set Point? ___—___   Auto/Manual? ___—___   Cascade? ___—___

Error Squared? ___—___   Error Deadband? ___—___

Gain = ___10___   Reset Time = ___999.___

Rate = ___0___   Reverse Acting? ___No___

Output Address: ___V219___   20% Offset? ___Yes___

ALARMS

Process Variable	Low = 200	High = 800
Deviation	Yellow = 100	Orange = 200

Alarm Specifications

The system provides a standard means of warning the operator if it detects that a process variable is out of safe operating limits as specified by the user. The data for the limits is given here.

Specific System Parameters

Table 6-1 shows the specifications for the gas flow control loop which is assigned loop number 2. The constants are stored in memory locations C200 through C214. The variables are stored in memory locations V200 through V217. The process variable has an offset, is in SCFH units with a maximum value of 1,000, is located at V57, and must have a square root calculation performed.

The gain for the control calculations is 10, the sample time 0.5 and the reset time 999.

Alarm points on the process variable are set at 200 and 800.

When this information is entered through the system programming units shown in *Figure 6-18*, the system is ready to operate, assuming that all electrical connections similar to those shown in *Figure 6-19* are made. Gain and integral times will have to be fine tuned from initial values to get the correct final system operation.

Continuous control also can be accomplished using a technique called time proportioning continuous control which will be discussed next.

TIME PROPORTIONING CONTINUOUS CONTROL

Instead of changing the amplitude of a manipulated variable in order to change a controlled variable, time proportioning continuous control varies the length of time a constant amplitude manipulated variable is applied.

Time proportioning continuous control means that the time that a controlled quantity is active or applied is varied rather than the amplitude of the quantity. It is commonly used in controlling electric heaters used in extruders, and in flow control applications by switching a metering pump on and off instead of controlling a pipe valve. It can be used in any application where the output control response is fast compared to the parameter being controlled; however, this requires an actuator that can be operated by being turned on and off rapidly and often.

VAT TEMPERATURE CONTROL EXAMPLE

Applications where time proportioning continuous control is very effective are those that use heat supplied by electric resistance heaters. Temperature control is accomplished by varying the amount of electricity that is supplied to the resistance heater element. In a proportional control system like the ones discussed up to now, this would be done by varying the voltage that is supplied to the heater element. However, with time proportioning control, the voltage remains constant and control is obtained by varying the amount of time that the voltage is applied to the heater element. The voltage is turned on and off at a rapid rate so that the average ON time determines the heat output.

As illustrated in *Figure 6-22*, the output of the control system is a pulse train with a varying duty cycle. The time from the start of a pulse to the start of the next pulse is a fixed period of time T. The pulse repetition rate is found by dividing one by T (1/T). If no output is allowed during this period, the duty cycle would be 0% and no heat would be produced. Conversely, if the output were on during all of the period, the duty cycle would be 100% and maximum heat would be produced. Normally, the duty cycle and the amount of heat produced will be somewhere in between these two extremes.

In this application, the temperature of the vat is to be maintained at 37°C (98.6°F). Since there is no cooling coil, the operation is such that if the temperature is over 37°C, the duty cycle is 0%, and at 27°C (80.6°F), the duty cycle is 100%. Therefore, the scale factor is 100% for 10°C or 10%/°C. The determination of the duty cycle is handled in the same manner as the standard proportional control; that is, the proportional, integral, and derivative controls are used in the required combination to determine the duty cycle to output to the heater element.

The width of the control pulse continuously varies the time the pulse is ON from 0% to 100% to control the temperature control

**Figure 6-22.
Time Proportioning
Control Output
Waveform**

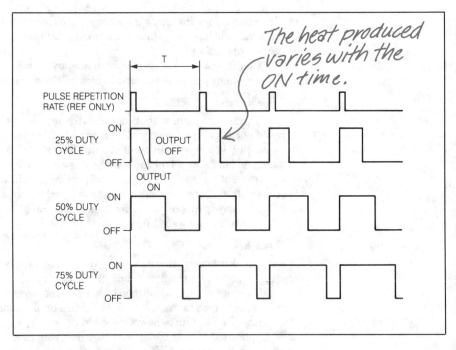

Using a PM550 as the Controller

A system for time proportioning control using a PM550 is shown in *Figure 6-23*. One PID control loop feeds its output to a time proportioning subsystem. The process variable is the vat temperature. The PID loop is formed by inputting to the system the gain, reset time and sample time. No other programming other than giving the range of the process variable (27 to 37) and the units (°C) is necessary to set up the loop except to provide the memory addresses for the constants and variables.

**Figure 6-23.
PM550 Block Diagram
for Time Proportion
Control**

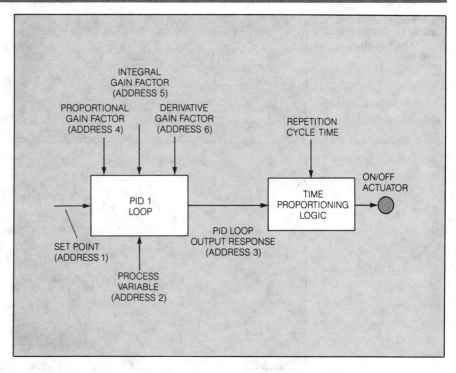

**Figure 6-23.
PM550 Block Diagram
for Time Proportion
Control**

One timer in the PM550
system determines the
repetition rate and the
turn-on of the ON control
pulse. A second timer controls the turn-off to control
the duty cycle.

The time proportioning output signal is formed by using standard
ladder logic functions, two internal timers and a special predefined function
contained within the PM550. The input to the time proportioning function is
the output from the PID loop which varies between 0% and 100% (0 and 1). A
diagram of the time proportioning subsystem is shown in *Figure 6-24*.

The first timer is an on-delay timer which determines the repetition
rate of the pulse. A normally-closed contact (C1) is used to reset and enable the
timer so that it cycles continuously. The special function uses the PID inputs to
calculate the count time for the second timer in the loop. The second counter is
then reset and enabled each time the special function completes its operation.
The second counter is enabled once each time the first counter times out since
the first timer is used to call the special function which reloads both counters.

SF8 is the predefined special math function in the PM550 that
performs the equation that follows the function call. Therefore, the instruction:

$$\text{SF8} \quad \text{V120} \times \text{C115} \div \text{C116} = \text{V121}$$

does the following:

It takes V120, which is the address of the result of the PID loop
response (a number between 0 and 1) and multiplies it times the repetition
period count preset for timer 1 (TMR1) stored in location C115. The product
of the values at V120 and C115 is divided by the value at address C116. The
contents of C116 is a conversion factor to ensure proper decimal point placement
of the result placed in address V121. The contents of V121 is the duty cycle count
which is the data used to load timer 2 (TMR 2).

**Figure 6-24.
Ladder Logic Diagram
for Time Proportion
Control**

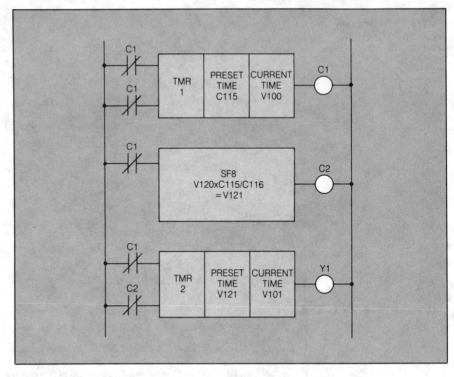

System Programming

The system is programmed by entering in the values listed in
Table 6-2 as the system prompts the programmer with questions as well as
programming in the special function instruction and the timer instructions
dictated by the ladder diagram.

One caution in using the time proportioning continuous control is
that not all processes have an output actuator which can respond fast enough
to be pulse driven. Thus, this type of proportional control has a more limited
application than the standard proportional control; however, it is compatible
with digital control since counting and time division are inherent in the way a
digital processor works.

LOOP RESPONSE DETERMINATION

Gain factors for such
time proportioned loops
usually are determined
experimentally.

The gain constants and update rate for each control loop must be
determined and, since the process is usually not known well enough to allow
the derivation of an equation which will describe it, the determination of
the gain factors is usually done experimentally. These need to be properly
specified so that the system produces the desired product and also to ensure
that the system will not have an unstable response. There are a number of
methods available to tune a control loop, one of which will be described here.
Typically, the system response is plotted versus time on a strip chart recorder
by manually inputting a disturbance and then looking at the response. One
method of determining the gain factors is described below.

Table 6-2.
Loop Specification Sheet

Loop Description: <u>Hot Water Vat Temperature Control Loop</u> PM550 Loop Number _____1_____

MEMORY ALLOCATION

Tuning Constants in <u>C</u> (C or V)

	Beginning Address	Table Length Tune in C	Table Length Tune in V	Ending Address
Constant Table	C100			
Variable Table	V100			

Are loop flags for alarms and mode switching allocated in the image register? No

If yes, give beginning address: —
(10 successive locations will be used)

PROCESS VARIABLE

20% Offset: No

Special calculation? No

Low Range = 27

Engr. Units °C

Address: A120

Square root? No

If yes, give address: —

High Range = 37

Transmitter —

CONTROL CALCULATIONS

Remote Set Point? No

Special Calculation? No

Lock-Set Point? No

Error Squared? No

Gain = 10%/°C

Rate = 0.0 min.

Output Address: V120

Sample Time = 10.0 sec.

If yes, give address: —

If yes, give address: —

Auto/Manual? Yes   Cascade? No

Error Deadband? No

Reset Time = 2 min.

Reverse Acting? No

20% Offset? No

ALARMS

Process Variable	Low = 75	High = 100
Deviation	Yellow = 4	Orange = 8

Ziegler-Nichols[1] Open-Loop Method for Determining Gain Factors

By introducing a known disturbance into a system and plotting its effect on the controlled variable, potential gain factors can be determined which will assist in evaluating and adjusting the control system.

This method was developed to determine initial setpoints for the gain factors used in process control loops. It is very effective at obtaining a satisfactory response characteristic without knowing a rigorous solution to the process response characteristics. A strip chart recorder is connected to monitor the system control variable as the system control is disturbed.

First the control loop is opened so that a manual disturbance (ΔM) can be introduced into the system — usually about 10%. The controlled variable is then plotted versus time. The output would look like that shown in *Figure 6-25*. A tangential line is then drawn through the inflection point of the control variable plot. The reaction lag (L_r) is the time from disturbance to the point where the tangent line crosses the controlled variable setpoint before the disturbance (base line). The reaction rate (R_r) is equal to the slope of the response curve at the inflection point. With this information, relationships have been developed which provide the ability to determine the gain factors for each type of control depending on the type of control loop. *Table 6-3* shows how the measurements made in *Figure 6-25* determine K_p, T_i and T_d. These values can be used to adjust or tune the system.

**Figure 6-25.
Step Response Provides
Data for Open-Loop
Tuning Methods**

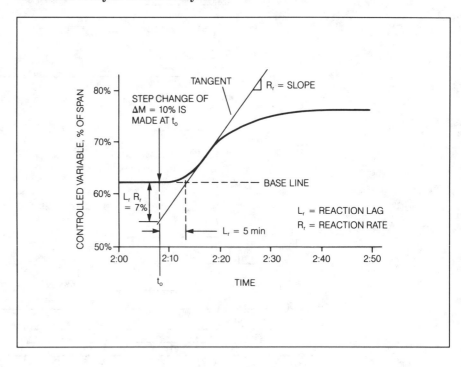

**Table 6-3.
Ziegler-Nichols Open-
Loop Gain Factors**

Proportional Only Control

$$K_p = \Delta M / L_r R_r$$

Proportional-Integral Control

$$K_p = 0.9 \, \Delta M / L_r R_r$$
$$T_i = 3.33 \, L_r$$

Proportional-Integral-Derivative Control

$$K_p = 1.2 \, \Delta M / L_r R_r$$
$$T_i = 2L_r$$
$$T_d = 0.5Lr$$

WHAT HAVE WE LEARNED?

1. Systems are controlled by proportional control, integral control, derivative control or a combination of proportional and any of the others.
2. Any of these control methods can use either analog or digital techniques.
3. Control systems can be partially analog and partially digital.
4. The error cannot be completely eliminated in a proportional-only system.
5. In an integral system, the error signal is summed over a period of time to produce the control.
6. In a derivative system, the control depends on the rate of change of the error signal with respect to time.
7. Microcomputer-based systems can provide all of the PID functions using digital techniques.
8. Microcomputer-based systems are extremely versatile and can use standard hardware to perform many different tasks, but they require a considerable amount of programming skills.
9. Microcomputer-based systems may require special software development equipment.
10. Programmable controllers are much easier to program. They prompt with questions to lead the programmer through the necessary input information required.
11. In a programmable controller, the PID loop control is already set up. Only the gain parameters, constants, input and output addresses, etc., need be supplied.
12. Microcomputer-based systems can be assembled easily by using preassembled boards that have been designed to operate together as a system.

Quiz for Chapter 6

1. What is a process variable?
   a. A controlled variable.
   b. A manipulated variable.
   c. A process parameter that changes value.
   d. All of the above.

2. What is process lag?
   a. The lag in the error signal when the setpoint is changed.
   b. The time it takes for the sensor to respond to a change.
   c. The time it takes for the mass of the process to respond to an input change.
   d. All of the above.

3. Most continuous control actuators:
   a. use a 4 to 20 mA current loop input.
   b. are of the time proportioning type.
   c. are used with standard continuous control loops.
   d. a and c above.
   e. all of the above.

4. Time proportioning control:
   a. can be used on a PM550 programmable process controller.
   b. cannot use a TM990 microcomputer.
   c. is less efficient than normal PID control.
   d. a and b above.
   e. all of the above.

5. What is proportional control?
   a. It has a discrete output for each error value.
   b. The controller output varies continuously over a range.
   c. The gain is independent of offset error.
   d. All of the above.

6. What is offset error?
   a. The difference between the sensor input and the absolute sensed parameter.
   b. The difference between the controller output and the setting of the manipulated variable.
   c. The difference between the control setpoint value and the actual value.

7. What is the advantage of PI control over proportional control?
   a. It allows lower proportional gain.
   b. It eliminates offset error.
   c. It can accommodate large load changes without accuracy loss.
   d. All of the above.

8. What is the difference between integral and derivative control?
   a. Integral control has slower response to an error signal.
   b. Derivative control does not eliminate offset error.
   c. Derivative control responds to a change in the rate of change in the error signal.
   d. All of the above.

9. The Ziegler-Nichols method:
   a. can be used to determine gain factors of a PI loop.
   b. is a graphical solution method.
   c. is a method of control.
   d. a and b above.
   e. all of the above.
   f. none of the above.

10. What two methods are available to build electronic continuous control?
    a. Pneumatic and analog.
    b. Analog and mechanical.
    c. Analog and digital.
    d. None of the above.

11. How is the integral mode implemented using analog methods?
    a. Op amp and resistors.
    b. Op amp and a capacitor.
    c. Op amp and an RC network to place the error voltage across the capacitor.

12. How is the integral mode implemented on a digital controller?
    a. The error is divided into discrete amounts.
    b. The digital controller cannot handle integral mode.
    c. The error signal is summed over a discrete time interval and accumulated by the controller.

Examples of Semi-Continuous (Jobshop) System Control

ABOUT THIS CHAPTER

Semi-continuous systems are those where a sequence of operations are performed. Each operation of the sequence is continuous (e.g., drilling a hole), but the operations as they follow each other are different (e.g., moving the part, using different size drill bit). The automated equipment may be programmable or non-programmable. If a large volume of the same part is to be manufactured over a period of years, the cheaper non-programmable machine may be chosen. However, for many applications, the programmable system is the best choice since retooling for the manufacture of a new part will be relatively inexpensive.

Programmable automation for semi-continuous parts manufacturing is called numerical control. This chapter describes numerical control systems and two innovations that have occurred, direct numerical control and computer numerical control.

NUMERICAL CONTROL

Drilling, milling and planing are semi-continuous machine operations that lend themselves well to numerical control (NC), programmed with APT language.

Numerical control (NC) is a method of controlling a semi-continuous parts manufacturing system by means of a set of instructions (a program) consisting of special arrangements of letters, numbers and symbols. The program is stored on paper tape, magnetic tape or other storage media and read by the machine each time that particular operation is performed. Many languages exist for writing an NC program, but the one most used is called Automatically Programmed Tools (APT). APT was originally developed in 1956 at MIT, but it has been modernized and expanded several times. The basics of APT were explained in Chapter 5.

TYPICAL NC OPERATIONS

If an operation involves production of parts made from similar feedstock (raw material) with variations in size and shape, NC is an appropriate choice. Even if production quantities are in small lots, NC can be economically feasible, but it is necessary that a sequence of operations be such that they can be performed on the same NC machine. However, for complete manufacturing of parts involving several sequences of operation that are dissimilar, several NC machines may be used.

Some of the operations that can be performed by NC machines include:

Planing	Quality Control	Milling
Drilling	Positioning	Extrusion
Shaping	Welding	Turning
Pressing	Riveting	Assembly
Cutting	Sanding	

Figures 7-1 and 7-2 show typical NC machines. The one shown in *Figure 7-2* is a CNC machine; this type is discussed later in this chapter.

**Figure 7-1.
Kearney & Trecker NC Machine**
(Photo courtesy of Kearney & Trecker Corporation)

**Figure 7-2.
Bridgeport CNC Machine**
(Photo supplied courtesy of Bridgeport Machines Division of Textron Inc.)

TYPICAL NC SYSTEM

A typical NC system flow is shown in *Figure 7-3*. The NC program is the set of instructions that is read by the machine controller. It is a step-by-step sequence to control the machine functions to manufacture the part.

Figure 7-3.
Typical NC System Flow

The punched tapes give the controller the operating instructions.

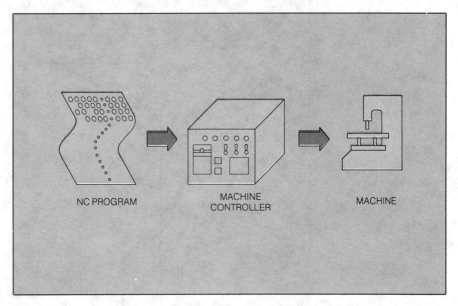

NC PROGRAM   MACHINE CONTROLLER   MACHINE

Program

A punched paper tape is the medium that provides the instructions to the machine. The holes in the tape represent a binary code that is input to a microprocessor in the controller.

The NC program is normally contained on a paper tape with each instruction being represented by a sequence of punched holes on the tape. The Electronic Industries Association (EIA) standard tape dimensions and format are given in *Figure 7-4*. Each row across the tape has eight positions and a hole is either present or absent in each position. One binary digit (bit) is represented by each position and eight bits (one row) represent one character (e.g., A, B, 2). A collection of one or more characters forms a word and a collection of words that forms a complete instruction is called a block. A collection of blocks (instructions) forms a complete machining process.

Controller

The controller is a critical component because it is the link between the NC program and the machine that actually performs the tasks. It reads an instruction from the paper tape by sensing the presence or absence of holes and translates it into a series of control signals. These signals are applied to the machine to cause its actions to perform the desired operations. A given controller will work with only one programming language and only one kind of machine.

**Figure 7-4.
Industry Standard NC
Punched Tape**

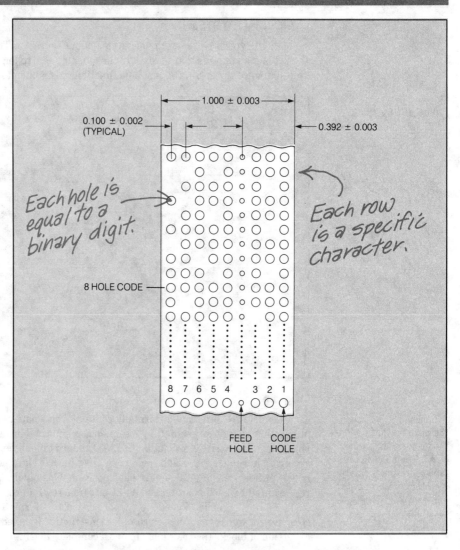

The controller reads, interprets the instructions and directs the machine to perform the operations desired. The machine operator is alerted when material must be inserted, tools changed, etc.

Some controllers have provisions for closed-loop operation so the machine returns a signal to the controller that indicates what the machine actually did. The controller can then compare the actual operation with the desired operation. If the error between these is significant, the controller can halt further operations and/or alert the operator either with a light signal that can be seen or a sound that can be heard or both. Adding this feedback to the controller is expensive and many of the newer systems have eliminated this feature because the machines are so accurate and dependable.

The controller also alerts the operator to functions that a human must perform such as loading feedstock, changing tools, or unloading the finished product. In this way, the operator does not need to continuously watch one machine and can perform other tasks.

Machine

The final component of an NC system is the machine itself. This component is normally called a machine *tool* because it performs some tooling process. However, all numerically controlled automation systems do not use a machine tool. An example is an NC system used to spray-paint labels on different parts of a product. The product is moved and label masks positioned to print the proper message at the appropriate locations on the product. The NC system controls movement of the product, selection and positioning of the label and activation of the spray gun. No machine tooling is involved.

In any case, the machine is the "money maker". It is the component that actually does the work. Its size and complexity is a function of the work to be done and the feedstock to be handled. It includes operator controls and often has a separate operator control panel for manual operation. The controllers for the controlled machine also are shown in *Figures 7-1* and *7-2*.

HOW TO USE NC SYSTEMS

The exact steps necessary to be followed in using an NC system is a function of the system chosen and the programming language used. In Chapter 5 when the APT language was discussed, a specific example of its use to machine a part was given. All systems, though, require the same thought processes:

1. Define the part.
2. Define the operations.
3. Develop the program.

HOW TO DEFINE THE PART

A three-dimensional grid is used for reference to give the machine its translational and rotational movement instructions and machining points.

The part is defined by dimensions so that the location of all operations to be performed are clearly specified. To do this requires a three-dimensional coordinate reference system to describe the movement of the work or the workpiece in reference to the coordinate system. In some operations such as drilling, the *workpiece* will be moved from hole location to hole location. In other operations such as turning on a lathe, the *work tool* will move along the workpiece.

If the NC machine has a table on which the part is mounted, the coordinate system is defined with respect to the table as shown in *Figure 7-5*. The x and y axes are across the surface of the table top for horizontal dimensions and the z-axis is perpendicular to the table top for vertical dimensions. (Note: The directions "horizontal" and "vertical" are used here with reference to *Figure 7-5*. Actually the coordinates could be rotated to any plane.) The origin of the axes does not have to be located at the corner of the table as shown, but an origin that is understood by the NC system must be defined. The movements in these three axes are called translational movements.

**Figure 7-5.
Coordinate System for
NC Machine**

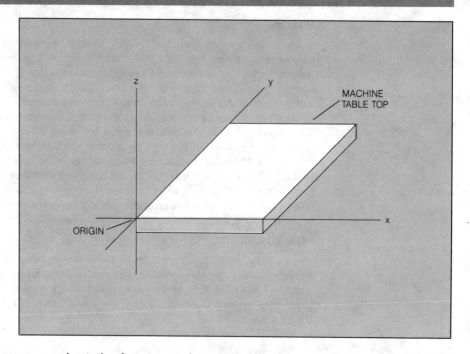

Each type of movement is called a "degree of freedom". Most systems have at least two degrees of freedom.

A rotational movement about each of the three axes also may be defined; therefore, up to three translational and three rotational movements can be used to describe the workpiece and its step-by-step positioning. Each movement is called a "degree of freedom" and NC systems often are rated as having n degrees of freedom where n can be 1, 2, 3, 4, 5 or 6. Most systems will have at least two degrees of freedom.

Examine the workpiece in *Figure 7-6*. All dimensions are given so that any point on the workpiece can be defined in reference to the origin of the coordinate system. The dimensions always are given in x, y, z order. Thus, point A would be described as (9.0, 6.0, 0.0) and point B as (6.0, 6.0, 3.0).

HOW TO DEFINE THE OPERATION

The starting point for a machine operation can be defined by absolute coordinates or by using coordinates that show relative position change to a previous operation.

The operation to be performed on a workpiece is defined by first specifying the starting and stopping point of the cutting tool using the coordinate system, then specifying the type of operation.

If the coordinate system is defined on the table top, the location of the tool in the "retracted" or "off" position is known. If a previous operation has just been completed, the coordinates of that location also are known. To move the tool to the start of the new operation, absolute coordinates with respect to the origin may be specified or coordinates relative to the last location may be specified.

**Figure 7-6.
Workpiece Dimensions
and Location**

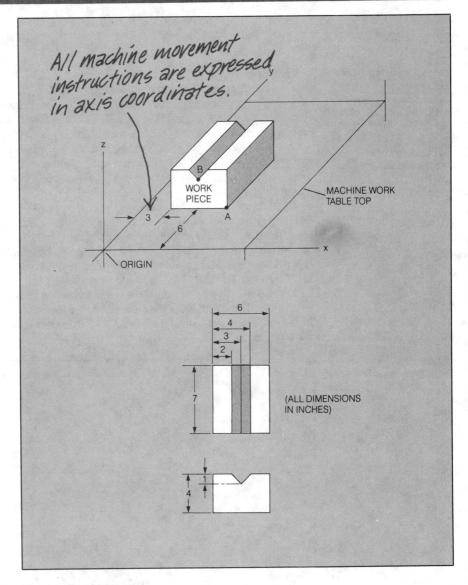

Point-to-Point Control

In *Figure 7-6*, assume that a cutting tool will be used to cut the V-channel out of the metal block in one pass of the tool. Also assume the cutting tool is currently located at (8.0, 18.0, 8.0). The bottom of the cutting tool is to be moved vertically to what will be the bottom of the V-channel and it is to be moved horizontally to a point one inch behind the block on the y axis and centered on the V-channel on the x axis. To specify the location in absolute coordinates, the following command is used.

GO TO (6.0, 14.0, 3.0)

To specify the location in coordinates relative to the current location of the tool, this command is used:

GO DELTA (-2.0, -4.0, -5.0)

Point-to-point control is the most basic type of machining control. Once it has been positioned at the starting point and given machining instructions the tool is moved from one specified coordinate to another.

which specifies a "delta" movement from the last location (8.0, 18.0, 8.0). The new location is calculated by the controller as (8.0 - 2.0, 18.0 - 4.0, 8.0 - 5.0) or (6.0, 14.0, 3.0), which is the same as the absolute coordinates in the GO TO command. This type of operation, called point-to-point control of the machine tool, is the most elementary control operation.

In this example, it is assumed that the workpiece is stationary and that the tool moves. The same procedure and commands work just as well if the tool is stationary in two axes and the workpiece moves. Of course, adjustment in coordinate values would need to be made.

A drill press is an example of the tool being fixed in the x, y directions, but moveable in the z direction. *Figure 7-7* shows a drill press table top with a 1 inch thick rectangular metal plate secured to it. Assume the drill bit (bottom of tool) is currenty located at (16.0, 2.0, 1.5). The drill bit is to drill a hole in the center of the plate which is located at absolute coordinates (10.0, 6.0, 1.0). Thus, GO TO (10.0, 6.0, 1.0) would locate the drill bit over the center of the hole. To use relative coordinates, the command would be GO DELTA (-6.0, +4.0, -0.5). If the drill bit is then moved by (-2.0) in the z axis, the bit will be forced into the metal to drill the hole. A (+2.0) in the z axis would then retract the bit from the metal.

**Figure 7-7.
Workpiece to be Drilled**

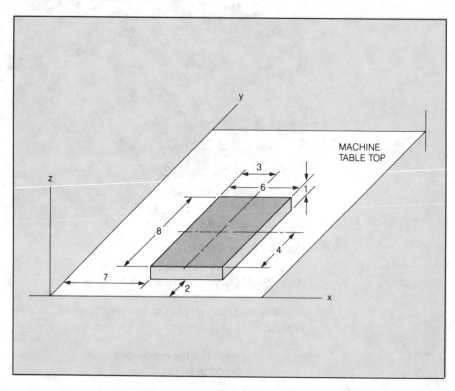

Straight-Line Control

Straight-line control is a type of point-to-point control where the cutting tool is moved parallel to one axis at a constant machining rate.

A second type of operation is straight-line cutting. Here the cutting tool is moved parallel to one of the three axes at a controlled rate (the feed rate) that is appropriate for the type of tool and material being used. Refer back to *Figure 7-6* and remember that the cutting tool was left in position ready to cut the V-channel. After power is applied and the tool is up to speed, the command GO TO (6.0, 5.0, 3.0) would feed the tool into the metal to cut the V and come to a stop one inch in front of the workpiece. Only the y axis coordinate was changed to accomplish this straight-line cut. Workpieces of rectangular shape are ideal for straight-line operations.

Contouring

When contouring, the machining operation must be capable of being defined in geometric terms, because the tool or the stock must be moved in two or three directions in one processing step.

The third operation, contouring, is considerably more complex than either point-to-point or straight-line operations. More than one axis or movement can be controlled at the same time so that movement can proceed in two or three directions at once. As with the straight-line operation, the feed rate must be controlled as a function of machine tool and material.

The simplest contouring movement would be a diagonal cut in the x-y plane across a workpiece as shown in *Figure 7-8*. A piece with a more complex contour, but still in two dimensions, is shown in *Figure 7-9*. For any contouring, the motion must be described geometrically or it cannot be done by NC. An even more complex three-dimensional contour to be machined in a workpiece is shown in *Figure 7-10*. Small dimensional steps programmed to machine the part in the three dimensions according to the geometrical pattern are required.

**Figure 7-8.
Simple Two-Dimensional
Contouring Operation**

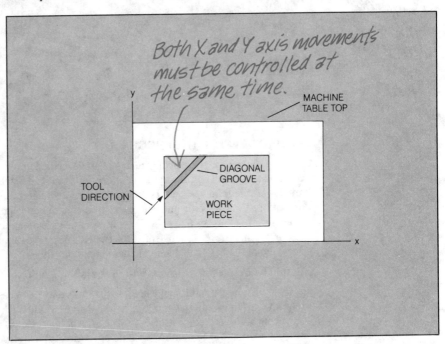

**Figure 7-9.
More Complex Two-
Dimensional Contour
Operation**

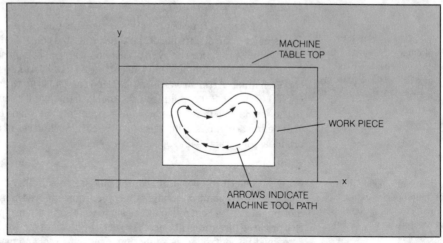

**Figure 7-10.
Complex Three-
Dimensional Contour
Operation**

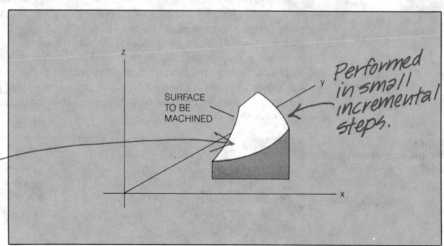

X,Y,Z coordinate movements must be controlled at the same time.

Performed in small incremental steps.

HOW TO DEVELOP THE PROGRAM

Now that the workpiece has been described in terms of a fixed coordinate system and the types of operations that the machine can perform are understood, how is a control program written to accomplish the operations? The specific answer to this question depends on the programming language. In Chapter 5, a specific example was given for the APT language. However, there is a sequence of steps that is common to almost all program development. Let's take a look at these common steps.

State The Sequence Number and Prepare the Machine

In the machining of a complex part, it may be necessary to change the tool several times and to change the type of operation. Each portion that can be performed with the same tool and same type of operation is called a sequence. It is given a sequence number and the operation type is stipulated.

For example:

```
n003
g00
```

where n003 indicates sequence number 3 (in this case a drilling operation) and g00 indicates a point-to-point operation.

Give The Coordinates

Regardless of the programming language, there are basic types of instructions that must be included beside the position information. Tool information, feed rate, cutting speed, machine on and machine off are examples of these basic instructions.

For a point-to-point operation, specify the coordinates of the point at which the operation is to occur. For straight-line and contouring operations, specify the coordinates of the beginning point of the operation. For straight-line, the ending point also is specified. For contouring, the geometry of the contour is described. Continuing with the example, the following gives the x and y coordinates for the center of a hole to be drilled:

```
x3.711
y - 0.960
```

Give the Tool Information

If a different tool is to be used, the tool number must be listed. Tools are usually numbered beginning with 1. For example,

```
t03
```

indicates to use tool number 3 in the NC machine. Let's assume it is a 0.375 inch drill bit.

Give the Feed Rate

The feed rate of the tool in inches per minute must be given. For the example instruction:

```
f60
```

f indicates feed rate and 60 specifies 60 inches per minute.

Give the Cutting Speed

The cutting speed is the rotational rate of the tool spindle in revolutions per minutes (RPM); it must be given for the cutting operation to be performed. For the example instruction:

```
s260
```

s indicates that it is an instruction for cutting speed and 260 is the RPM.

State Any Miscellaneous Requirements

This is where special instructions such as start machine, wait for operator, or change tools can be given. The form is:

```
m07
```

where m indicates a miscellaneous instruction and 07 designates the specific instruction to be followed. Let's assume in this case it is to turn on the drill bit coolant.

Putting it Together

Combining the sequence of operations:

```
.
.
.
n003
g00
x3.711
y − 0.960
t03
f60
s260
m07
n004
.
.
.
```

Sequences n001 and n002 would precede n003, and n004 would follow if additional operations were required. Sequence 3 causes a hole to be drilled at location (3.711, -0.960) with a 0.375 inch drill bit. The hole will be drilled at a feed rate of 60 inches per minute with a spindle speed of 260 rpm. The coolant will be turned on during the drilling operation.

COMPUTER NUMERICAL CONTROL

Computer numerical control systems provide the advantages and flexibility of electronic memory to semi-continuous operations, and provide virtually wear-free operation.

As discussed earlier in this chapter, a punched paper tape containing the NC program is generated for a particular task. This tape must be read for every workpiece that is made using the program. If production is 1,000 units per week, the tape is read 1,000 times per week. Of course, duplicate tapes are made so that a worn tape can be replaced, but the tape and mechanical tape reader are the most common sources of problems in NC systems. In computer numerical control (CNC), a computer memory rather than tape is used to store the program and the program is read electronically. Thus, there are no moving parts or tapes to cause a malfunction. The computer is dedicated to the machine it controls and usually is located next to it.

A typical CNC system diagram is shown in *Figure 7-11*. The original program input still may be from paper tape, or it may be from magnetic tape, or it may be input directly from a keyboard. After the program is input or loaded from the external source, it is stored in the computer's RAM. It doesn't ever need to be re-entered until it is changed unless power is lost to the computer. On older model machines, the computer may be a minicomputer, but new CNC systems almost always use microcomputers because of size and cost reductions. The interface unit provides signal conditioning to change the computer output signals to the type of signals required by the machine as discussed in Chapters 3 and 4.

**Figure 7-11.
A CNC System**

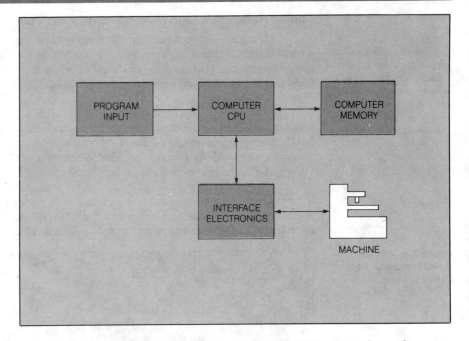

Computer controlled systems can handle more complex contouring operations than punched tape NC controllers, as well as provide adaptive control and monitor the process for quality and proper operation.

Besides the advantage of electronic memory, CNC systems have other advantages over traditional NC systems. Setup or development time of a new process is reduced because programming corrections and program optimization can be accomplished easily on the CNC system. This programmability also increases productivity of the machine because it is easy to change the program and make adjustments to optimize the machine's movements.

Because of the computer's capability for mathematical manipulation, more complex contouring can be accomplished. Because of the computer's ability to monitor sensors, closed-loop control is easier, and elements of the process can be continuously monitored and varied for best results. For instance, the feed rate of a drill bit may be increased until the drill motor slows too much or the temperature of the drill bit exceeds a predetermined level, then the feed rate would be slowed. Such a method would allow holes to be drilled at the fastest rate possible without burning bits or overloading the drill motor. This is called adaptive control. The result of applying adaptive control is increased productivity.

Another advantage of CNC is that the computer can monitor itself and the machine. Diagnostic programs for detecting or locating electronic or mechanical failures can be contained in the computer memory. These can be set up to run automatically or under operator control. As a result, the machine itself is locating and identifying the cause of troubles or failures. Still another advantage of CNC is the further use of the computer's calculating and graphing ability. For example, some statistical functions such as productivity versus time of day can be obtained and easily outputted by the computer. A final advantage of CNC is its compatibility with direct numerical control.

DIRECT NUMERICAL CONTROL—THE HIGHEST PLANE

The advantages of centralized operational control are found in direct numerical control (DNC). A large central computer connected to the individual controllers located at each machine is the heart of the DNC system.

If a small computer can directly control one machine as in CNC, it seems reasonable that a large computer can control several machines. In fact, this can be done and the resulting system is called direct numerical control (DNC). One DNC computer can control up to 256 machines, but most DNC systems control fewer.

Historically, DNC arrived on the scene before CNC. The older DNC systems used hardware controllers rather than computers at each machine, but their programs were loaded directly from the memory of a large central computer instead of from paper tape. The development of CNC fit right into DNC systems and offered immediate improvements.

The central computer in DNC is located remotely from the machines it controls and is connected to them by wires called telecommunications lines. *Figure 7-12* shows a possible DNC layout. The DNC computer supervises the operation of all the machines in real time. It selects and loads machine control

**Figure 7-12.
Direct Numerical Control
System**

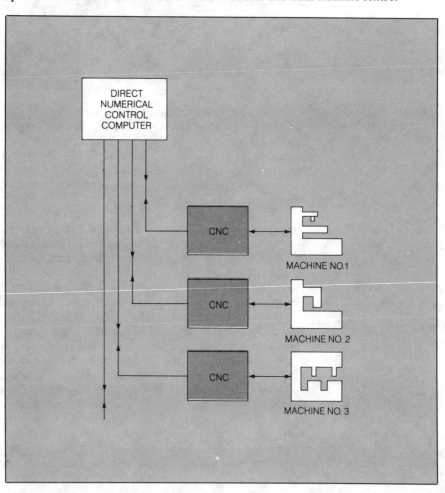

programs into the CNCs. It also receives machine status reports and requests for service. An operator at each CNC can ask questions of the DNC computer through the CNC computer and receive answers and instructions. The DNC computer can store data concerning the operation and productivity of all operators and machines and use the data for reporting and analysis purposes. A big disadvantage of DNC is that the entire operation may be shut down if the DNC computer fails.

WHAT HAVE WE LEARNED?

1. Automatically controlling a semi-continuous parts manufacturing system by a program of instructions is called numerical control.
2. The main components of an NC system are the program, the controller and the machine.
3. The development of the program requires defining the workpiece within a given coordinate system and selecting point-to-point, straight-line or contour operations. These choices are limited by the machine capabilities.
4. If the program is contained on a tape that must be read for every operation by a non-computerized controller, the result is a standard NC system.
5. If the controller is replaced by a computer and the program is stored in memory, the result is a CNC system.
6. If a central computer is used to control more than one machine, the result is a DNC system.
7. Increased productivity and decreased production costs are the result of properly using the NC, CNC or DNC concept in manufacturing involving semi-continuous operatons.

Quiz for Chapter 7

1. A semi–continuous operation means:
   a. coffee breaks are allowed for the operator.
   b. the operation is continuous each shift, but stops between shifts.
   c. a computer is a required part of the machine.
   d. each operation in a sequence is continuous, but subsequent operations in the sequence may be different.

2. Numerical Control:
   a. is a method for controlling the operation of a machine by means of a set of instructions.
   b. applies only to milling machines.
   c. is a method for producing an exact number of parts per hour.
   d. none of the above.

3. A typical NC system has which of the following components?
   a. tape input.
   b. controller.
   c. machine tool.
   d. all of the above.

4. The coordinate system used to describe a workpiece:
   a. changes from part to part.
   b. is always two-dimensional.
   c. is defined with respect to a point on the machine.
   d. all of the above.

5 Straight-line operations:
   a. must be limited to two per workpiece.
   b. must be parallel to one of the axes of the coordinate system.
   c. can include diagonal cuts if they aren't curved.
   d. all of the above.

6. Which of the following is not specified in a computer program?
   a. cutting speed.
   b. feed rate.
   c. time required to make the piece.
   d. type of operation.

7. A disadvantage of Numerical Control (NC) is:
   a. the computer is not reliable.
   b. the tape and tape reader are not reliable.
   c. the machine tool can easily overheat.
   d. one operator is needed for each machine.

8. Direct Numerical Control means:
   a. using several computers to control one large machine.
   b. using one computer to control several machines.
   c. eliminating computers from the manufacturing process.
   d. complete elimination of a need for a human operator.

9. In an NC instruction; if the tool is to move from location (10, 10, 10) toward zero two units in the x direction, away from zero four units in the y direction, and toward zero 6 units in the z direction; the instruction would be:
   a. GO DELTA (-2.0, -4.0, $+6.0$)
   b. GO DELTA (-2.0, $+4.0$, -6.0)
   c. GO DELTA ($+2.0$, -4.0, $+6.0$)
   d. GO DELTA (-2.0, -4.0, -6.0)

10. In question 9, the final position of the tool after execution of the instruction is:
    a. (8, 14, 4)
    b. (12, 6, 16)
    c. (12, 14, 16)
    d. (8, 6, 4)

Discrete Parts Manufacture Using Programmable Controllers

ABOUT THIS CHAPTER

This chapter deals with the use of programmable controllers for discrete parts manufacturing. The first part of this chapter explains the programmable controller operation in relation to relay logic, then actual applications of a PC are discussed along with the explanation of a controlled machine's operation. A programmable controller (PC) was selected as the controller for these applications only after careful consideration of the advantages and disadvantages of the PC. Some of the reasons for using a PC are given and the things that had to be considered in selecting a particular PC are discussed. In all cases, the flexibility, reliability, and ease of troubleshooting of the PC provided significant advantages over other types of control. A summary and explanation of the important parameters and specifications of a PC are given to aid in evaluating PCs.

PROGRAMMABLE CONTROL INTRODUCTION

Many discrete parts manufacturing processes involve a sequential operation. This made possible the application of relay logic, followed later by the programmable controller.

Most manufacturing processes require a sequence of operations in order to build a product. This is particularly true of discrete parts manufacturing. The sequencing can be done either manually or with some type of controller. Up until the late 1960's, sequencing of discrete manufacturing operations and many other manufacturing operations usually was performed using a bank of relays uniquely wired to perform the particular task. Thus, the use of relay logic is known very well in most industries. However, since relay logic may be difficult to troubleshoot and modify, the need for a more reliable and standardized system was apparent. These facts, along with the availability of semiconductor logic functions, resulted in the development of the electronic programmable controller. Since relay logic was so well known, the controller design engineers developed the programmable controller so that the same "language" and ladder diagrams of relay logic still could be used. This allowed the technicians to program and use the programmable controller with very little retraining.

Ladder Diagram Explanation

In a ladder diagram, the vertical rails are the power source, and the circuits are the rungs. Contacts are represented by a pair of short parallel lines. Inputs are X, outputs Y, and control relays, C.

Figure 8-1a shows a motor connected to a power source through a switch. *Figure 8-1b* shows the equivalent ladder diagram. In a ladder diagram, the power source is represented by the two vertical "rails" of the ladder and the various control circuits make up the "rungs". Normally open contacts of a switch or relay are symbolized by two parallel vertical lines as shown in *Figure 8-1b*. (Don't confuse this with the similar capacitor symbol used on schematic diagrams for electronic circuits.) Normally closed contacts are symbolized by the parallel vertical lines with a diagonal line across them. Inputs are designated by X and control relays are designated by C. Outputs are represented by a circle and a Y designation.

**Figure 8-1.
Motor Control Diagrams**

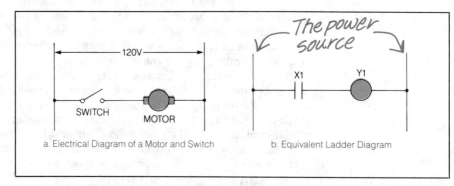

a. Electrical Diagram of a Motor and Switch   b. Equivalent Ladder Diagram

Switch or relay contacts can be wired into the various logic functions shown, OR, AND, and combinations of the two.

Figure 8-2 shows other ways the motor can be controlled. Consider X1, X2, X3 and X4 as contacts of switches that are shown in the OFF position of the switch. In a, if either switch X1 or X2 is turned ON the X1 or X2 contacts close and the motor is energized. This is called the OR condition. In b, both switches X1 and X2 must be ON so both contacts X1 and X2 are closed to energize the motor. This is called the AND condition. In c, with the switch X4 OFF, the motor is energized through a set of normally closed contacts (X4) on the switch. When X4 is switched ON, its normally closed contacts open and the motor can be energized only by turning ON X1 or X3 to close either the contacts X1 or the contacts X3. If both of the switches X2 and X4 are turned ON their normally closed contacts will be open, then turning ON X3 to close contacts X3 is the only way to energize the motor.

Control Relay

A control relay is shown schematically in *Figure 8-3a*. Besides providing decision-making through logic circuit connections, a relay is used rather than a switch to control power at a remote location and/or to control a high-voltage or high-current device with a low-voltage low-current switch. With control switch X1 open as shown, relay C1 is de-energized and output Y1 is connected to the 120V source through the normally closed (N.C.) contacts of C1. When X1 is closed, the coil of C1 is energized and the movable contact moves to close the normally open (N.O.) contacts. Now power is removed from Y1 and applied to Y2. Notice that all parts of the same relay are designated by the same C number in the equivalent ladder diagram in *Figure 8-3b*.

Figure 8-2.
Ladder Diagrams for
More Complex Ways to
Control a Motor

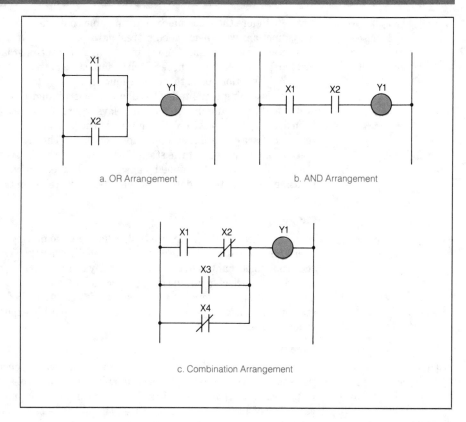

a. OR Arrangement

b. AND Arrangement

c. Combination Arrangement

Figure 8-3.
Diagrams for Relay
Controlled Devices

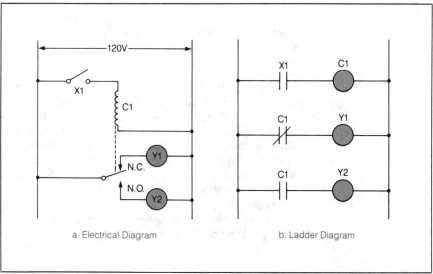

a. Electrical Diagram

b. Ladder Diagram

In PC's, state-of-the-art technology utilizes solid-state memory to control logic functions and actual relays only for power control to actuators.

Before the electronic programmable controller was developed, the relay functions were performed by real hardware relays with physical wiring connected between them. When a programmable controller is used, the only real hardware relays are those required for actual outputs to the machine. Any relay logic functions used to store a sequence state or to combine multiple relay logic paths are implemented with solid-state circuit digital logic referred to in design and programming as "control relays" because real relays would have been used in the hardwired relay method. These "control relays" are not real relays, but are only solid-state memory locations. The digital code contents of these locations represent the state of a "relay contact". Therefore, the state of these "contacts" is determined by examining the memory contents rather than by using a test meter to determine whether a real relay contact is set open or closed.

Safety

Since the PC hardware is all solid-state (no moving parts) and the operation of the control logic occurs in memory rather than with actual relays, the problem of changes in the timing of relay contact opening or closing is eliminated. Such changes in timing, which are common in real relays, may change machine sequencing and cause unsafe operation. In contrast, a solid-state memory is very reliable and has a much longer life expectancy than a relay contact.

Timers

The functions of electromechanical timers, counters and relays can be duplicated effectively by the programmable controller.

Electronic equivalents of electromechanical sequencing drum timers, time-out timers and counters also are built into the controller and need only to be programmed to be used. No additional hardware is required. This extra capability of a PC often can simplify and improve the performance of the control system without increasing the cost.

In summary, PC controllers perform the same function as electromechanical relays, timers and counters; can perform mathematical calculations, can be easily programmed rather than rewired; and have a performance that is much more reliable.

Intelligent Peripherals

Although this discussion is concerned only with the implementation of digital functions using the PC, the line separating the programmable controller from the continuous analog type is becoming less distinct. For instance, it is possible to add an "intelligent" I/O peripheral unit to the sequencer type of PC. Not only is the intelligent unit programmable, but it also may incorporate an analog I/O capability. These peripherals are simply turned on and off by the main sequencer and then the peripherals perform their operation independently of the main unit. The intelligent units can handle such tasks as stepping motor control, serial communication interfaces, and analog PID control if needed.

The programmable controller operates in sequence. It accepts the input signals from sensors of outside systems status, does only the mathematical manipulations required and sets the output actuators to provide the action required.

**Figure 8-4.
PC Functional Block Diagram**

HOW A PROGRAMMABLE CONTROLLER WORKS

The programmable controller basically acts as a programmable sequencer when used to automate a machine. It first senses the input conditions of interest, then solves the logic equations programmed by the control engineer based on the current input conditions. It then sets the required output actuators to provide action as dictated by the solution of the logic equations. The block diagram of a PC in *Figure 8-4* shows the four main functional blocks. Let's suppose it is used to control a machine.

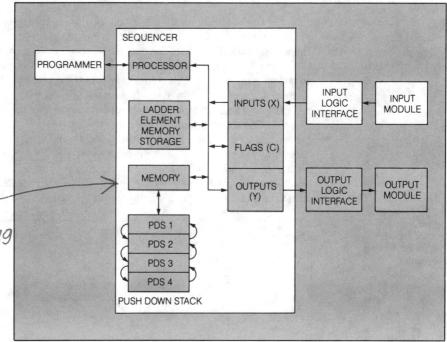

Programmable solid state logic performs timing, latching and control instead of hardware components.

Input

Any external parameters needed for system control come to the PC through the inputs.

The input block allows the sequencer to sense the machine input parameters. These may consist of operator switches used to set up and select modes of machine operation and for starting or stopping the machine operation, limit switches used to sense position or presence of a part, and other sensors or switches used for safety or fault monitoring. In short, any external parameter which must be provided to the controller for it to make the proper decision when running a machine is provided through the input block.

Sequencer

The sequencer provides the timing, temporary storage and overall system control using solid-state logic.

The sequencer block performs the same operations of timing, latching and control that hardwired relay logic does, except the operations are done with programmable solid-state logic rather than with discrete hardwired components. The sequence logic is basically digital codes using the same digital logic signals previously described.

Programmer

The programmer is used to enter the program that the PC uses for system control.

The sequence is controlled by a user-defined program that contains the steps necessary to execute the task from the ladder diagram which is equivalent to the relay logic. The user program is entered through the programmer block. The sequencer itself has its own operating program — sometimes referred to as a microprogram — which causes the sequencer hardware to step through the user program and execute it according to the predefined rules of the system.

Output

The action required for controlling the process or machine is accomplished through the outputs.

In the output block, the output points are turned on or off according to the results of the ladder logic based on inputs at any particular time. Remember that the state of the outputs is stored in memory. This means that the on or off state of these outputs can be used within other ladder logic equations simply by the program examining the contents of the appropriate memory location. If hardwired relays were used, each logic equation would require using a separate set of contacts for a particular output.

The actual sequence of reading the input, solving the control sequence based on the input conditions, and then setting the outputs accordingly is done on a fixed time cycle determined by the external power line frequency as shown in *Figure 8-5*. The scan time at 60 Hz is 8.3 milliseconds per 1024 words (1K) of memory. At 50 Hz, the scan time is 10 milliseconds. The scan time increases 8.3 (or 10) milliseconds for each additional 1K of memory.

**Figure 8-5.
Typical I/O and
Sequence Timing
Diagram**

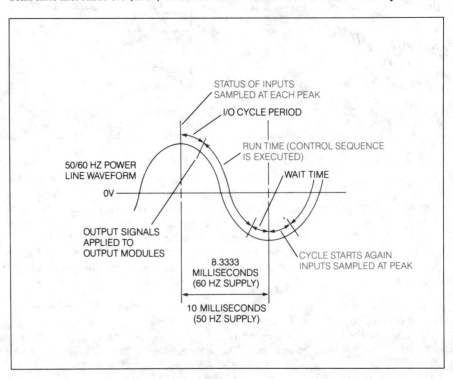

BAG HANDLER CONTROL USING A PC

An example of a small machine which uses a programmable controller is a bag handler[1] used to fill bags with fertilizer, animal feed, or other types of dry-fill bulk products. The basic machine operation is straightforward. Bags must be added to a bag-supply hopper in the machine as needed by an operator. The rest of the operation can be totally automatic or can be controlled manually if the operator so chooses. The bags are drawn from the bag hopper one at a time and placed under a fill hopper. A previously weighed batch of material is dispensed from the fill hopper into the bag. Then the bags are dropped onto a conveyor belt for transfer to the sealing station.

Choosing A Controller

The bag handler system lends itself to programmable controller operations.

This machine has less than ten moving parts; however, the parts are used for millions of cycles per year in a dusty uncontrolled environment. In the past, the control of this machine might have been handled with a relay bank and some timers. In this application, the type of PC considered and the total number of I/O points were such that the hardware costs of relays versus a PC were almost identical. Typically, the cost of a PC will be about the same as a relay implementation if about two-thirds or more of the I/O capability of the PC is used. Even if only a few relays are needed, the PC still may be a better choice even if the cost is more because of its flexibility, reliability, and ease of use. This allows the machine development time to be shorter but, more importantly, the recurring machine checkout and troubleshooting times will be much shorter which will result in significant time savings and a better satisfied user.

For the bag handler, the control engineers decided to develop a local controller rather than rely on a central system controller. The local controller was designed to operate either in an automatic mode or a manual mode. This provided localized troubleshooting and, at the same time, resulted in an easier design.

Machine Design

Besides choosing the type of controller, a number of alternatives faced the development engineers. For example, they had to decide how fast the machine must operate and the range of bag sizes allowed. In addition, the type and location of power and other plant facilities had to be considered.

General considerations for a bag handling system include: the speed of operation, the various bag sizes to be handled, the type power to be used, the installation, etc.

In the final design, vacuum cups were chosen to pick the bag from the stack and pull it up around the fill hopper. Due to the heavy weight of a filled bag, a set of mechanical clamps was used to clamp the bag around the hopper during filling. Also, since different length bags were used at various times, a motor driven roller mechanism was used in the bag hopper to push the bottom part of a long bag out of the hopper. This was necessary since the bag pick arms did not move far enough to completely withdraw a long bag. By using this method, the physical size of the machine did not have to be increased to handle long bags.

[1]Inglett & Co.

Two manipulator arms were chosen: One to pick the bag from the stack with vacuum cups and bring it to the hopper; the other, also using vacuum cups, to pull it up around the hopper. It was necessary to design controls for each of these arms and for the mechanical clamp that supports the weight of the filled bag. The double-acting pneumatic cylinders that move the arms and the mechanical clamp operate from air pressure which is controlled by the PC through solenoid valves. Solenoid valves also are used by the PC to control the vacuum to the cups that hold the bags. A pressure relief valve on the clamp cylinder allows pressure to be removed quickly to permit a quick bag release after filling. A line drawing of the machine depicting the control points referenced to inputs and outputs is shown in *Figure 8-6a*. The input/output (I/O) assignments are shown in *Figure 8-6b*.

I/O Assignments and Wiring Considerations

The number of I/O points required is one of the parameters that must be known to choose the controller hardware for the application. *Figure 8-6b* shows a total of 16 points; 8 outputs and 8 inputs. For this application, a Texas Instruments Model 510, which has 20 I/O points, proved both adequate and cost-effective.

With the monitoring and control points and the controller defined, the actual input and output connection terminals on the controller need to be specified so that the controller can be programmed and the machine control wiring diagram can be drawn. Since the control program does not yet exist, no consideration need be given to it when selecting the I/O connection points. Even if the program did exist, usually it is easier to change the I/O assignment in the program than the physical wire connections to the controller. As a result, the I/O assignments shown in *Figure 8-6b* were made to make wiring easier rather than to gain any control logic advantage.

All inputs and outputs must be checked to determine the voltage and current requirements to make sure that when the I/O terminals are wired to the sensors and actuators that there is sufficient drive voltage and current for the actuators and proper power supplied to the sensors.

Control System Development

Once the machine is defined by performance and application specifications, a control program can be developed. The first step is to determine the sequence of operations of the machine and to identify the appropriate input and output control points to accomplish that sequence. The basic overall sequences for totally automatic operation and for manual operation are shown in *Figure 8-7*.

**Figure 8-6.
Bag Handler I/O Control**

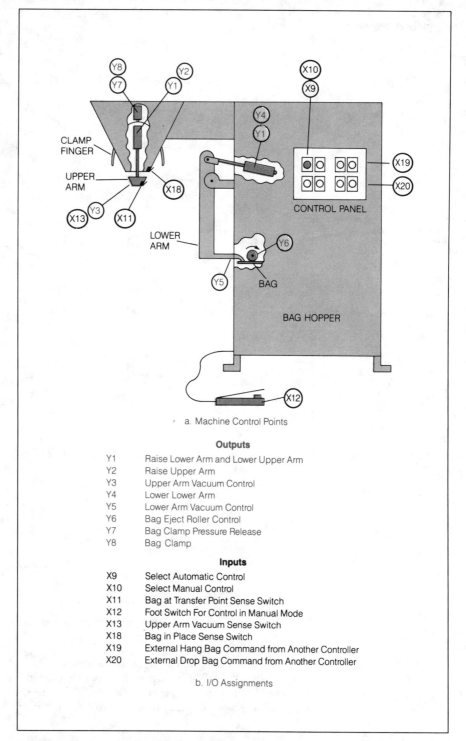

a. Machine Control Points

Outputs

Y1	Raise Lower Arm and Lower Upper Arm
Y2	Raise Upper Arm
Y3	Upper Arm Vacuum Control
Y4	Lower Lower Arm
Y5	Lower Arm Vacuum Control
Y6	Bag Eject Roller Control
Y7	Bag Clamp Pressure Release
Y8	Bag Clamp

Inputs

X9	Select Automatic Control
X10	Select Manual Control
X11	Bag at Transfer Point Sense Switch
X12	Foot Switch For Control in Manual Mode
X13	Upper Arm Vacuum Sense Switch
X18	Bag in Place Sense Switch
X19	External Hang Bag Command from Another Controller
X20	External Drop Bag Command from Another Controller

b. I/O Assignments

**Figure 8-7.
Simplified Flow Diagram**

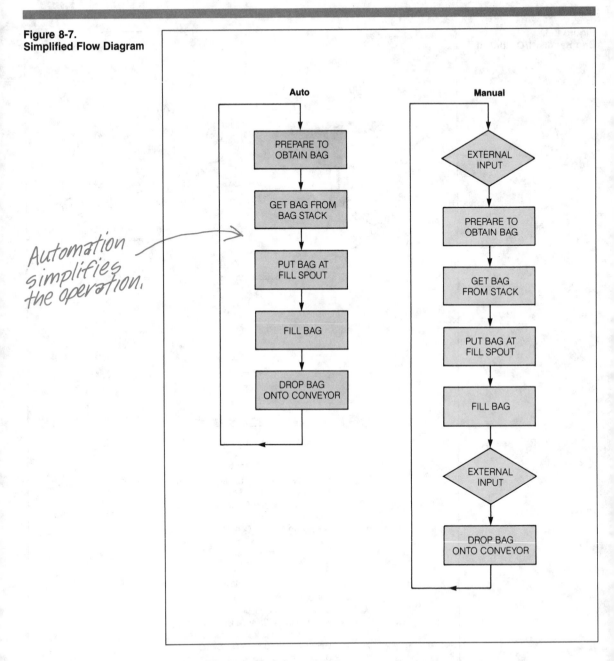

Automation simplifies the operation.

Ladder Diagram Development

The flow diagram in *Figure 8-7* is expanded to that in *Figure 8-8* which has more detail. The relay logic ladder diagram in *Figure 8-9* is developed from the flow diagram in *Figure 8-8*. During the following discussion of this ladder diagram, it may be helpful to refer to *Figures 8-7 and 8-8*.

**Figure 8-8.
Bag Handler Flow
Diagram**

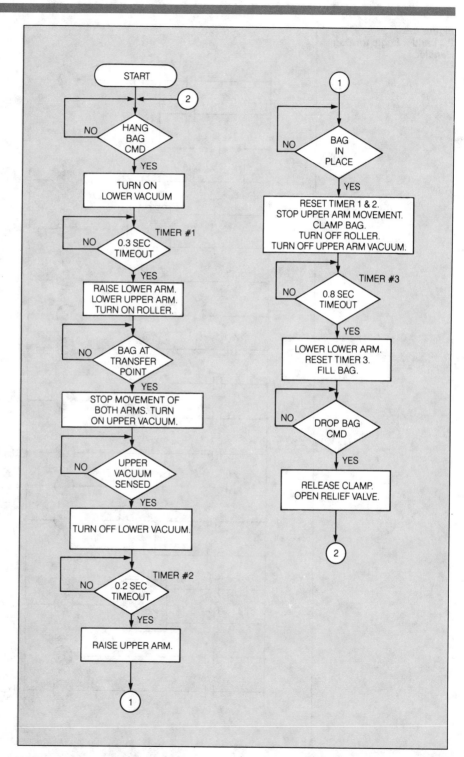

**Figure 8-9.
PC Ladder Logic for Bag
Handler**

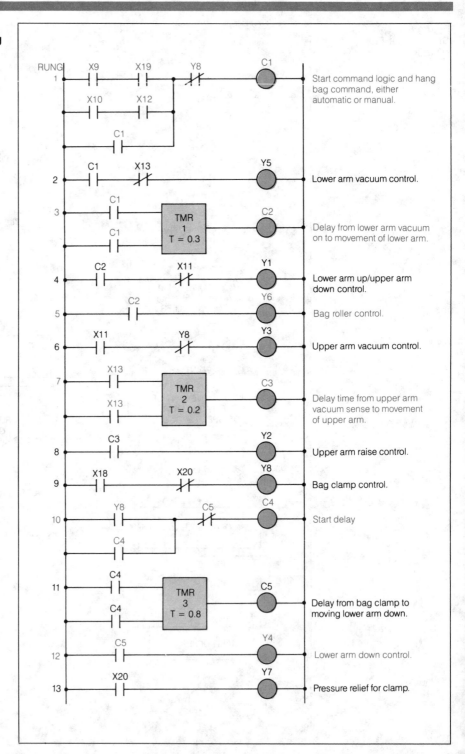

RUNG
1 — Start command logic and hang bag command, either automatic or manual.

2 — Lower arm vacuum control.

3 — Delay from lower arm vacuum on to movement of lower arm.

4 — Lower arm up/upper arm down control.

5 — Bag roller control.

6 — Upper arm vacuum control.

7 — Delay time from upper arm vacuum sense to movement of upper arm.

8 — Upper arm raise control.

9 — Bag clamp control.

10 — Start delay

11 — Delay from bag clamp to moving lower arm down.

12 — Lower arm down control.

13 — Pressure relief for clamp.

Automatic bag handling and filling is achieved by a series of command-initiated sequential operations.

a. The bag is picked from a stack.
b. The bag is positioned and clamped around the hopper for filling.
c. The bag is filled.
d. The full bag is dropped on a conveyor.

In rung 1, either automatic operation (X9) or manual operation (X10) will be in effect as selected by the operator. A hang bag command from another controller (X19) in automatic, or a hang bag command from the operator's foot switch (X12) in manual, energizes control relay C1 which self latches through the C1 input. In rung 2, when C1 closes, the lower arm vacuum is turned on to pick a bag from the stack through solenoid valve Y5. In rung 3, when C1 closes, timer 1 is started and times out in 0.3 second.

C2 is then energized which energizes Y1 in rung 4. Y1 turns on the air pressure to the air cylinder actuator which raises the lower arm to bring the bag up to the fill hopper and also to the air cylinder actuator that lowers the upper arm to the bag transfer point. C2 also energizes Y6 in rung 5 to turn on the bag roller motor to eject the bottom of the bag from the hopper.

When the bag reaches the transfer point, a sensing switch (X11) is tripped by the physical presence of the bag. This opens Y1 to stop movement of both arms. X11 is held on by the bag until it is dropped. In rung 6, when X11 closes, Y3 is energized to turn on the upper arm vacuum so that the bag is grabbed by the upper arm. This vacuum is sensed by switch X13 whose open contact in rung 2 turns off Y5 to turn off the lower arm vacuum so the bag is released from the lower arm. In rung 7, the closed contacts of X13 start timer 2 which times out in 0.2 second, then energizes C3. In rung 8, C3 energizes soleniod valve Y2 to raise the upper arms. The V shape of the fill hopper pushes the bag open as the upper arm pulls the bag up around it.

When the bag is in place as sensed by switch X18 in rung 9, Y8 activates the mechanical bag clamp. Also Y8 releases C1 in rung 1 to reset timer 1 which de-energizes Y6 in rung 5. Y8 also releases Y3 in rung 6 to turn off the upper arm vacuum. With the upper arm vacuum off, X13 opens and resets timer 2. This de-energizes Y2 to stop upward movement of the upper arm. In rung 10, Y8 energizes C4 which self latches and starts timer 3 in rung 11. It times out in 0.8 second and energizes C5. In rung 12, C5 activates Y4 which turns on air pressure to the lower arm cylinder to move the arm down to pick another bag. In rung 10, C5 opens to release C4 which resets timer 3.

A premeasured quantity of material is released through the fill hopper into the bag under control of another controller. Then this same controller inputs a drop bag command (X20). In rung 9, X20 opens to release the bag clamp solenoid valve Y8 and, simultaneously in rung 13, energizes solenoid valve Y7. Y7 opens a pressure relief on the clamp air cylinder so the clamp is released very quickly from the bag and the bag drops onto a conveyor. The bag handler controller then waits for the next hang bag command to repeat the cycle.

Programming the Controller

The logic for the first rung is entered into the controller using the handheld programmer in the following sequence (remember that these program instructions were discussed in Chapter 5):

```
STR X9 ENT
AND X19 ENT
OR X10 ENT
AND X12 ENT
OR C1 ENT
AND NOT Y8 ENT
OUT C1 ENT
```

where ENT is the ENTER key that causes the instruction to be stored in memory.

Additional rungs in the ladder diagram are programmed in the same fashion using the commands of Chapter 5. For example, rung 10 is:

```
STR Y8 ENT
OR C4 ENT
AND NOT C5 ENT
OUT C4 ENT
```

and the timer TMR 3 is:

```
STR C4 ENT
STR C4 ENT
TMR 3 ENT
48 ENT
OUT C5 ENT
```

The count number for TMR 3 is 48 because 48×16.67 milliseconds $= 0.8$ second, the required delay time for TMR 3. Some programming debugging may be necessary to get the timers set properly, but with a small amount of effort the machine is set for operating.

TRIM AND BORE MACHINE CONTROL USING A PROGRAMMABLE CONTROLLER

A manufacturing process of trimming a piece of wood and drilling holes in the ends can be performed under the control of a PC unit.

The woodworking industry also uses automatic machine control to perform the many tasks that are repeated over and over again. One common task is to trim a piece of wood to a set length and then drill holes in each end to a set depth. This process requires the proper sequencing of the saws and drills to avoid conflicts with one another and to ensure that the process is done safely and correctly. The sequence of the operations can be provided using a variety of control methods; three types that have been used are relay, mechanical, and programmable controller logic. The PC has proved to be both cost effective and reliable in this application. It improved reliability in the dusty environment and reduced troubleshooting time. It also has been more flexible and provides better safety protection than the other types of control.

Machine Operation

The basic operational sequence to be controlled is as follows:

a. An operator inserts the piece of wood.

b. The start switch is pressed.

c. Two circular saws are lowered to cut the ends, then raised.

d. Two drill presses insert drills into the ends and drill the holes, then retracted. Limit switches control the depth.

e. The machine stops and the piece removed.

The operation of the double-end trim and bore machine shown in *Figure 2-4* is straightforward. A wood piece is inserted into the machine by the operator and the operator starts the machine by depressing a start switch. First, the piece is clamped to prevent slipping. The machine then trims both ends at the same time by lowering two circular saws into the wood until they close limit switches at the end of travel. The closed limit switches cause the saws to be raised to their home position. Then two drills bore holes in both ends of the wood at the same time to a preset depth. When they close limit switches at the end of their travel, they retract to their home position. When all operations are finished, the clamp holding the piece is released so the finished piece can be removed.

At the present time, the wood piece is inserted and removed manually, but these operations could be performed by a loader mechanism under control of the same PC. An air cylinder is used for the clamp and hydraulic cylinders are used to move the saws and drills. These cylinders are controlled by solenoid valves which in turn are controlled by the PC.

A total of 20 input/output points are required - eight outputs for the saws, drills and clamps and twelve inputs for monitoring limit switches and operator switches for the operating modes. The required operating speed can be handled easily by the TI Model 510 which was selected as the PC. The I/O assignments shown in *Figure 8-10* were used to interconnect the output actuators and input switches and for programming the controller.

**Figure 8-10.
Trim and Bore Machine
I/O Assignments**

Outputs		Inputs	
Y1	Clamp	X9	Pressure Loss Monitor Switch
Y2	Left Saw Extend	X10	Left/Right Saw at Home
Y3	Right Saw Extend	X11	Left Saw Extended
Y4	Both Saws Retract	X12	Right Saw Extended
Y5	Left Drill Retract	X13	Left/right Drill at Home
Y6	Left Drill Extend	X14	Left Drill Extended
Y7	Right Drill Retract	X15	Right Drill Extended
Y8	Right Drill Extend	X16	Cycle Start
		X17	Select Left Drill
		X18	Select Right Drill
		X19	Select Left Saw
		X20	Select Right Saw

Sequential Control Using Drum Timer

Since the machine operation repeats the same tasks in a sequence over and over again, one of the easiest ways to control the sequence is with a drum timer. The programmable controller provides a drum timer function electronically so that an external physical drum timer is not required.

The timing diagram depicting the state of each output is shown in *Figure 8-11*.

Figure 8-11.
Trim and Bore Timing
Diagram

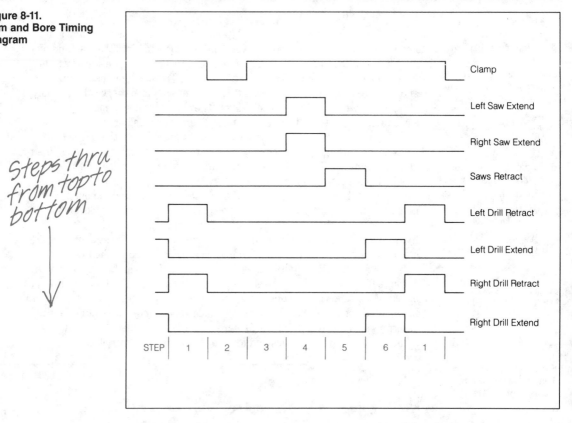

Steps thru from top to bottom

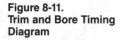

As shown in the timing diagram, there are six unique sequential steps.

The timing diagram is divided into steps where each step produces a unique set of output conditions. Since there are six unique conditions, the controller must have six unique states to sequence the machine properly. The standard drum timer in the Model 510 PC used in this application has fifteen unique outputs and sixteen unique steps. Since only six are needed, the unused ones must be programmed to zero so that no action is required or performed as the timer goes through steps 7 through 16.

The drum timer function allows the use of 15 logic outputs that act like control relays. These control relay outputs are combined with the switch inputs to provide the eight unique states shown in the timing diagram of *Figure 8-11* for each of the six steps.

In addition, some of the control relay outputs are used for the stepping sequence control. The programming of the drum timer is shown in *Figure 8-12*. Note that control relays C12 through C15 (read numbers vertically) are programmed to count in binary from 1 to 15. By combining the input and output logic from the controller with these control relay outputs, the drum timer will not advance to the next state until all actions have been completed. As a result, the drum timer function makes the controller self-sequencing. Note that the dwell time at each step can be controlled by the counts-per-step (CNT/STP in *Figure 8-12*) programmed into the controller.

**Figure 8-12.
Drum Timer Program for
the Trim and Bore
Machine**

DRUM 01 PRESET = 01 SCN/CNT = 01	C 0 1	C 0 2	C 0 3	C 0 4	C 0 5	C 0 6	C 0 7	C 0 8	C 0 9	C 1 0	C 1 1	C 1 2	C 1 3	C 1 4	C 1 5
STEP CNT/STP*															
1 00001	1	0	0	1	1	0	1	0	0	0	0	0	0	0	1
2 00001	0	0	0	0	0	0	0	0	0	0	0	0	0	1	0
3 00060	1	0	0	0	0	0	0	0	0	0	0	0	0	1	1
4 00001	1	1	1	0	0	0	0	0	0	0	0	0	1	0	0
5 00001	1	0	0	1	0	0	0	0	0	0	0	0	1	0	1
6 00001	1	0	0	0	0	1	0	1	0	0	0	0	1	1	0
7 00000	0	0	0	0	0	0	0	0	0	0	0	0	1	1	1
8 00000	0	0	0	0	0	0	0	0	0	0	0	1	0	0	0
9 00000	0	0	0	0	0	0	0	0	0	0	0	1	0	0	1
10 00000	0	0	0	0	0	0	0	0	0	0	0	1	0	1	0
11 00000	0	0	0	0	0	0	0	0	0	0	0	1	0	1	1
12 00000	0	0	0	0	0	0	0	0	0	0	0	1	1	0	0
13 00000	0	0	0	0	0	0	0	0	0	0	0	1	1	0	1
14 00000	0	0	0	0	0	0	0	0	0	0	0	1	1	1	0
15 00000	0	0	0	0	0	0	0	0	0	0	0	1	1	1	1
16 00000	0	0	0	0	0	0	0	0	0	0	0	1	1	1	1

*COUNTS/STEP INDICATES LENGTH OF TIME OUTPUTS ARE IN THE PRESCRIBED STATE
FOR EACH STEP.
1 - INDICATES OUTPUT WILL BE ON DURING STEP.
0 - INDICATES OUTPUT WILL BE OFF.

Drum Timer Ladder Logic

The ladder diagram is divided into the six unique steps.

The ladder logic diagram for the state sequencing is shown in *Figure 8-13*. The first rung sets up the enable and run lines of the drum timer and the remaining rungs set up the sequence control of the drum once it is running. The first line of the first rung indicates that the drum is stepped by control relay C30 which is the output of the drum event logic (state sequencing). This allows the use of the event logic to step the drum because C30 is active only when all conditions of each step are true. The second line of the first rung indicates the conditions that will enable the drum to operate. It is enabled when C31 (the drum control relay) AND C29 (the error detection relay) AND X9 (the air pressure switch) indicate that there is no error condition and no loss of air pressure while the drum is running; otherwise, this rung will reset the drum to the beginning state. The drum number, its preset state, and the step count must be given to completely specify the drum. (Each of these is 01 as shown in the upper left corner of *Figure 8-12*.)

**Figure 8-13.
Drum Timer Ladder
Logic**

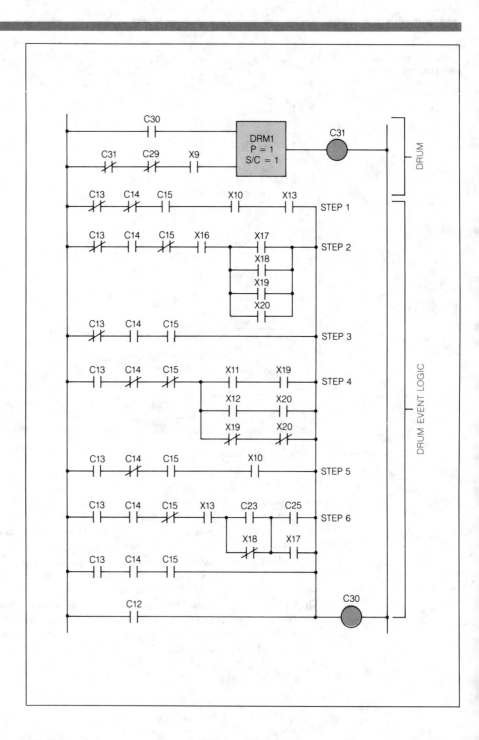

The program for the Texas Instruments 510 controller to accomplish the logic of the drum rung of the ladder diagram in *Figure 8-13* is as follows:

```
STR NOT C31 ENT       These four lines set up the parameters for C31,
AND NOT C29 ENT       which is the reset enable line of the drum timer.
AND X9 ENT
OUT C31 ENT

STR C30 ENT           This sequence sets up the drum timer itself.
STR C31 ENT
DRM 1 ENT
1 ENT
1 ENT
OUT C31 ENT
```

Drum Event Ladder Logic

The drum timer is operated in sequence so that conditions in step 1 must be met before the drum timer moves on to step 2 and so forth through step 6.

The sequence senses whether the saws and drills are at the home position before continuing. The saws are turned on, travel to their limit position and then are returned to home.

The next sequence of ladder logic in *Figure 8-13*, the drum event logic, describes the operation of the drum to provide the steps and states in the timing diagram discussed above. Let's briefly review the operation. All the step inputs and outputs are ORed to control relay C30 which is the drum count enable relay. It must set (the "contacts" must be closed) in order for the drum to continue to count. C12 through C15 are drum control relays which are set depending on the step count of the drum itself. Since these also are used in each step of the ladder logic, the remaining logic in that step must be true during that step count for the drum to continue to operate.

In the first step of the drum event logic, X10 (indicates saws are at home position) and X13 (indicates drills are at home position) must be true before the timer will go to the next step. These conditions are included in step 1 to ensure the saws and drills are fully retracted before the machine can start. This is a safety feature to protect both the machine and the operator. Step 2 requires that X16, the start switch, be activated by the operator before the timer will advance, else the timer will stay in this rung and wait. When the start signal is given, X17 through X20, which are the operator switches to select which drills and saws to run on the next cycle, are checked. If none are selected, no machine action is required and the controller stays in step 2. If any one is selected, the timer moves to step 3. Step 3 is simply used as a delay and no external logic is required. The step time count is set at 60 as shown in *Figure 8-12* which results in a 0.2 second delay on the Model 510 controller. After the 0.2 second delay, the timer advances to step 4.

Step *4*, according to the drum timer program in *Figure 8-12*, sets control relays C1, C2, and C3 which activate the clamp and saws. The saws will run and move until they reach the travel limit established by limit switches sensed by inputs X11 and X12, then the timer advances to step 5. In step 5, control relay C4 in the drum program activates the raise mechanism to return the saws to the home position. When the saws are fully retracted, the limit switch makes X10 true and the timer moves to step 6.

A similar sequence occurs for the drills.

Step 6 causes the drills to run as commanded by control relays C6 and C8 in the drum program. C6 is for the left drill and C8 is for the right drill. The drum sequencer stays at step 6 until the drills reach their limits. The remainder of the drum timer is not used so it steps through steps 7 through 16 without any action, then resets to step 1. At step 1, the drills are commanded home. The timer then moves to step 2 to wait for operator action to start the next cycle.

Output Ladder Logic

The ladder logic to produce the outputs that operate the machine takes only eight rungs as shown in *Figure 8-14*. This gives some indication of the power of the PC controller with the drum sequencer in this application.

Figure 8-14.
Output Ladder Logic

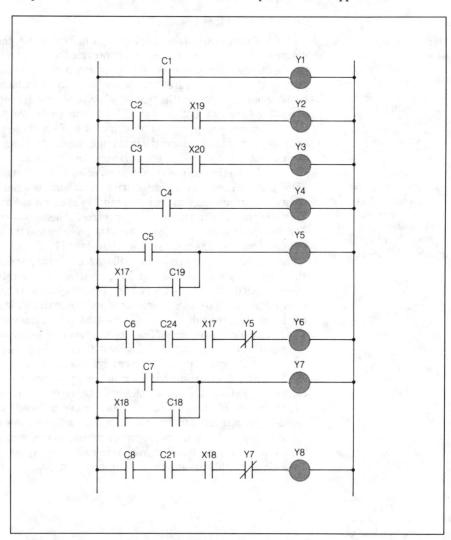

Let's go through this logic in the order of the output actuator
assignments. Y1, which is the air-operated clamp that holds the wood piece in
place, is controlled by C1 only. The drum timer program in *Figure 8-12* shows
that control relay C1 is on for the entire cycle after the start switch is operated
in step 2. Thus, the wood piece is clamped before any other action and stays
clamped until the drills retract in step 1. Y2, the left saw actuator will be on
only in step 4 and only if X19, the select left saw switch is on. Likewise, the
right saw actuator, Y3, will be on only in step 4 and only if X20, the select right
saw switch is on. Y4, which raises both saws, is activated by C4 in step 5 after
the saws have completed the run in step 4. Y6 and Y8 extend the drills selected
by X17 and X18, respectively when C6 and C8, respectively, are true in step 6,
after the saws have retracted. Including Y5 and Y7 in the logic for Y6 and Y8,
respectively, makes sure that the drill-extend actuator is energized only if the
drill-retract actuator is not on. This prevents the machine from trying to
extend and retract the drill at the same time, a situation that would put
unnecessary stress on the machine. Finally Y5 and Y7, which are the drill
retract actuators, are active when C5 and C7, respectively, are true in step 1
and when the limit switch has been activated as indicated by C19 and C18,
respectively.

Error Detection Ladder Logic

The error detection logic determines if the machine is malfunctioning.
If so, all of the outputs are disabled to prevent equipment damage. An option
allows an error lamp to be programmed to turn on to notify the operator.

The error logic used in this machine consists of the four relay logic
ladder rungs shown in *Figure 8-15*. The first rung shows that when the
normally closed contacts of C27 are open, the master control relay MCR1 will
disable all of the machine output lines. The machine outputs will remain
disabled until an operator corrects the problem and restarts the machine. In
the second rung, when Y4 (both saws raised) AND Y5 (left drill retracted)
AND Y7 (right drill retracted) are true, then a timer (TMR1) starts to count
down for a three second delay. When it reaches zero, it activates timeout
control relay C28 which is used in the fourth rung in the shutdown logic. The
third rung checks for malfunctions in the drills' and saws' limit switches. If
both the home limit switch and fully extended limit switches are on at the same
time, C29 is activated. In the fourth rung, if C28 is on AND either C29 is on
OR X9 (pressure sense switch) is not open, then the control relay C27 is
activated and latched. Therefore, when the machine cycle is complete, if the
monitored error conditions exist or if air pressure is lost, control relay C27 is
latched on.

**Figure 8-15.
Error Detection Logic**

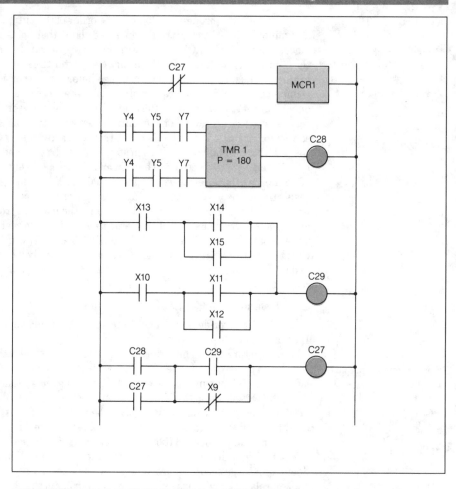

EVALUATION OF A PROGRAMMABLE CONTROLLER

The speed of operation of a PC system may be an important system parameter for many applications.

When choosing a PC, there are a number of characteristics that are important to consider. Usually these are specified in the manufacturer's literature; however, care must be taken when comparing specifications since the meaning can vary from one manufacturer to another.

Scan Time

One of the most important parameters, especially for a time critical operation, is the speed with which the PC can execute the ladder logic in the program. This is usually specified as the scan time per 1,000 nodes of logic. The nominal range of today's controllers is from 1 to 200 milliseconds. However, the meaning of this number must be considered carefully for a particular application because this time usually does not include the actual time spent in the loop routines that are associated with inputs and/or outputs. Such routines can add considerable execution time, thus, the controller speed may be much slower than the scan time calculated from the specification.

Synchronous Versus Asynchronous Operation

In synchronous operation, the I/O routines occur in series with the solution to the ladder sequence. In asynchronous operation, the I/O routine occurs independent of the ladder sequence.

Programmable controllers vary in how they handle sensing inputs and setting outputs (I/O routine) in time relation to the sequencing through the ladder logic. If the I/O routines occur in series with the solution of the ladder logic sequence, the operation is termed synchronous. If the I/O routines and sequencing through the ladder logic are occurring independently, the operation is termed asynchronous.

Asynchronous operation can appear to be faster and have better throughput than synchronous depending on how the operating loops mesh together in time. A problem with the asynchronous system, however, is that input conditions may change at any time in the sequence scan. For instance, suppose the same input connection is used at two points within a scan, one at the beginning and one at the end. If the input changes during the scan, then the output will not be consistent for a given cycle. This can lead to apparent intermittent logic errors where one or more outputs may have the incorrect value for a short period. This can be a serious problem in some machine controls, especially if the outputs are fast acting and can respond quickly compared to the speed of the controller. When evaluating a PC for a design, both the average throughput and worst case throughput should be considered.

Number of Inputs and Outputs

A basic specification for a PC is the total number of inputs and outputs that can be connected to the controller. The controller used should have enough I/O to meet the present application with some spares to allow for future expansion or improvements. Also of interest is whether the PC has I/O expansion capability. If I/O expansion units can be added efficiently and effectively, this will save overdesign in the initial unit and allow easy expansion as the system grows.

I/O Electrical Capability

Each input and output is capable of either accepting or sensing (sinking) or supplying (driving) a given amount of current or voltage. This capability must be compared to the items that will be attached to the controller to make sure that they are compatible. It is possible to save external hardware if the controller can provide enough voltage or current to drive an external device directly rather than having to add hardware to adapt between the controller and the device.

User Memory

The amount of user memory required is related to the length and difficulty of the logic that must be solved in an application. Simple applications with just a few control relays do not require a significant amount of memory. Of course, more complicated applications will require more memory. An estimate of the program logic is necessary to determine how much memory should be purchased. Programs for systems tend to expand after the system is designed so the ability to add more memory usually will be an advantage.

Matrix Functions

Matrix operations are used primarily for error detection. They allow the user to compare a set of conditions whose outputs are known with the actual machine input and output conditions to determine if an error has occurred. If so, a recommended corrective action can be defined based on the results of the comparison. This capability can reduce troubleshooting time by notifying the operator of the problem and recommending corrective action.

Arithmetic Functions

Unlike relay logic, a programmable controller can do arithmetic functions which can be used to set up timers and modify data. This increases the capability of the control system and results in better machine control.

Timers and Counters

With relay logic, timers and counters must be implemented with separate control elements which adds to the amount of hardware as well as the cost of the system. With the PC, these functions are handled in the program (software) and don't increase the cost of the system. Furthermore, the PC functions are more accurate and are limited in number only by the available program memory.

Program Development Support Hardware

Advances are being made in PC peripherals to display ladder logic programs as they are being developed, to store the program on magnetic tape, or to print out the program to have a hard copy.

Programming ladder logic is greatly simplified if it can be displayed as it is being developed and can be changed easily as the checkout progresses. If the PC has the capability to store the program on magnetic tape or other non-volatile media, then it does not have to be re-entered if power is lost. A CRT display and magnetic disk storage fulfills these needs and such units are available with some programmable controllers. Others have audio cassette tape units and handheld terminals similar to calculators. Another requirement for program development is a hard copy (on paper) of the program to allow proper documentation. Thus, a printer interface on the programmer unit usually is needed.

One other item is very important — service and support. Even if the operation of the controller itself meets all the application requirements, it may not be very useful if the proper development support is not available or if the proper service for repair and maintenance is not available.

WHAT HAVE WE LEARNED?

1. Discrete manufacturing operations that require something to be done over and over again in a certain sequence are particularly suited for control by a programmable controller.
2. Programming a programmable controller is easier if a ladder diagram is developed first.
3. A relay logic diagram can be converted to a ladder diagram.
4. A programmable controller first senses input conditions, then uses these input conditions to solve logic equations programmed into its memory and sets outputs determined by the results of the logic equations.
5. Flow diagrams are important aids for programming a PC.
6. Each rung of a ladder diagram can be programmed into the PC as a distinct portion of the program.
7. Programmable controllers have counter and timer capabilities and may even simulate a drum timer function.
8. Programmable controllers also can have error detection logic programmed into them.
9. Important characteristics of PCs that should be considered when choosing one for a system are:
   a. Scan time.
   b. Synchronous versus asynchronous operations.
   c. Number of available inputs and outputs and expansion capability.
   d. Electrical characteristics of inputs and outputs.
   e. Memory size and expansion capability.
   f. Matrix functions.
   g. Program development support from the manufacturer.
   h. Maintenance and repair service from the manufacturer.

Quiz for Chapter 8

1. Programmable controllers:
   a. are programmed using relay ladder logic.
   b. cannot perform functions other than relay logic.
   c. are poor industrial controllers with limited use.
   d. are widely used in many different industries.
   e. a and d above.

2. The PC is composed of:
   a. input modules.
   b. control sequencer.
   c. ladder memory.
   d. output modules.
   e. all of the above.

3. Relay ladder logic is used in a PC because:
   a. it is the best industrial control language ever developed.
   b. it is impossible to write a control program for a machine in any other logic.
   c. most industrial control personnel are familiar with relay ladder logic diagrams and this is the easiest for them to use.
   d. an international committee chose relay logic as the standard logic for industrial control.
   e. none of the above.
   f. all of the above.

4. Today, programmable controllers:
   a. are relatively low cost compared with the original digital equipment available for industrial control.
   b. are more powerful then the first controllers on the market.
   c. can be cost effective even when replacing just a few relays.
   d. none of the above.
   e. all of the above.

5. One advantage of a PC in industrial control is that it:
   a. has a reliable long life compared to relay logic.
   b. is very difficult for the inexperienced programmer to understand.

   c. is unreliable because there are no moving parts.
   d. none of the above.

6. A programmable controller:
   a. can be used on more than one machine during its lifetime.
   b. can be reprogrammed in the field.
   c. requires considerable training before it can be used.
   d. a and b above.
   e. none of the above.

7. In the operation of a programmable controller:
   a. all of the logic status is maintained in memory.
   b. mathematics cannot be performed.
   c. any latched data must be done using an external latching relay.
   d. all of the above.
   e. none of the above.

8. A programmable controller:
   a. allows faster machine checkout.
   b. is easier to troubleshoot than standard relay logic.
   c. all of the above.
   d. none of the above.

9. A ladder rung in a PC program:
   a. must contain at least one output.
   b. can contain AND, OR, NOT entries.
   c. are continuously scanned and the appropriate output set or reset.
   d. all of the above.
   e. none of the above.

10. The two types of relays in a PC are:
    a. an output relay and a control relay.
    b. a control relay and a retentive relay.
    c. a retentive relay and a non-retentive relay
    d. none of the above.

A New Dimension — Robots

ABOUT THIS CHAPTER

An important new tool for increasing productivity through electronic control of automation systems is the robot. The robot has characteristics to improve, or even make possible, the automation of certain production and assembly tasks. However, it probably would not even exist had it not been for electronic control. The use of robots in industrial applications in the United States was compared to other countries (particularly Japan) in Chapter 1. This comparison showed that the U.S. is far behind. However, interest now is keen in the U.S. and the development of smaller, less expensive robots with more complex motion and vision systems is allowing robots to be used in applications not possible before. This chapter defines a robot, describes several types of robots, and discusses when they are beneficial.

WHAT IS A ROBOT?

Industrial robots must be capable of being programmed, multifunctional in their use and be able to manipulate an object or a thing according to instructions.

The mention of robots normally produces visions of characters such as the droids *R2-D2*® and *C-3PO*® from the movie *STAR WARS*®. Indeed, *R2-D2* and *C-3PO* belong to a specialized class of robots, but they are not the type of robot presently used in industrial applications.

One of the dictionary definitions of "robot" is so broad that the automatic garage door opener, the electric dishwasher, or the refrigerator could be called a robot. However, the Robot Institute of America has a more refined definition:

"A robot is a reprogrammable, multifunctional manipulator designed to move material, parts, tools, or specialized devices through variable programmed motions for the performance of a variety of tasks."

First, the robot must be programmable; that is, one must be able to tell it what to do for a variety of circumstances that may occur. Further, it must be able to be reprogrammed to perform a different task. (This rules out the garage door opener.) Second, the robot is capable of performing more than one task — it's multifunctional. (This eliminates the dishwasher). Next, it must be a manipulator; that is, it must use force to move a workload from point-to-point or along a path step-by-step as instructed by a program. (This eliminates the refrigerator.)

INDUSTRIAL ROBOT CATEGORIES

The two basic categories of industrial robots are the pick-and-place manipulator and the intelligent robot.

Pick-and-Place Manipulator

The number of different directions in which a robot can move are termed its "axes of movement."

The pick-and-place manipulator is a low- or medium-technology device capable of point-to-point operation. *Figure 9-1* shows a typical pick-and-place robot. This particular robot has three axes of movement because it can move in the vertical plane, and the horizontal plane, and can rotate the gripper. Thus, the term "axes of movement" refers to how many ways the robot can move. Two other axes of movement are available as options. The maximum payload of this robot is 1.65 pounds (including the gripper).

Figure 9-1.
A Typical Pick-and-Place Robot
(Courtesy of Seiko Instruments USA, Inc.)

Pick-and-place manipulators are one type of robot used in small parts assembly/handling; typically operating in from three to five "axes of movement."

This type of robot is used primarily in small parts assembly, small parts handling, and at stations that feed materials to a process. Pick-and-place robots are capable of placing items with good accuracy and can repeat that placement with good accuracy (the Seiko 200 accuracy is ± 0.0004 inch). These robots are very reliable and can be re-used for other applications. Control of program sequencing for the Seiko 200 has been accomplished by the use of a 5TI or TI510 programmable controller.

Intelligent Robot

The intelligent robot has a memory and some are capable of independent decision-making. These robots are classified as medium- or high-technology devices depending on their level of intelligence.

Medium-Technology Robot

Medium-technology robots can be "taught" the operation they are to perform by following a human teacher through the moves required.

The robot shown in *Figure 9-2* is an example of a medium-technology intelligent robot used for spray painting. A skilled human paint sprayer moves the robot through the required motions and these are recorded, usually on cassette tape, for future use. This is referred to as "teaching" the robot by "walk-through." Once the teaching is completed, the tape is played and the robot repeatedly reproduces this motion. The paint gun mounted on the robot has a maximum speed of 5.6 feet/second and items up to 81 inches high and 36 inches deep can be painted. This particular robot has six axes of movement.

Figure 9-2.
A Medium-Technology
Intelligent Robot
(Courtesy of Binks Manufacturing Company)

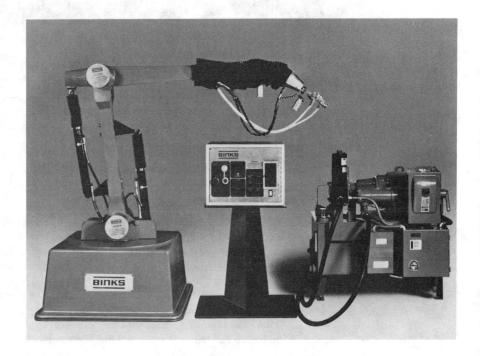

High-Technology Robot

High-technology robots have sophisticated computer controls and are extremely flexible in their capabilities and applications.

High-technology robots have highly sophisticated decision-making capability and are capable of performing complex tasks. *Figure 9-3* shows an example of a robot classified as a high-technology robot. This robot can handle a payload of 100 pounds with a positioning accuracy of ± 0.05 inch per axis. It has six axes of movement and can achieve speeds of 50 inches per second. It has a sophisticated computer control that allows it to branch to different areas of a program so that parts of its "standard" sequence of operations can be deleted, changed or added to. Its work envelope is 1,000 cubic feet. Some examples of uses of such a high-technology robot are given below.

**Figure 9-3.
High-Technology Robot**
*(Courtesy of Cincinnati
Millacron)*

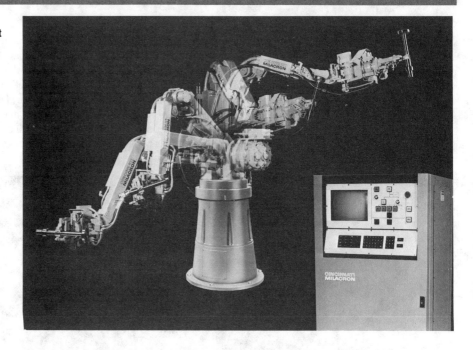

Welding: A 100 pound welding gun spot welds auto bodies that are moving on an assembly line at variable speeds up to 10 inches per second. If the line slows or stops, the robot can accommodate for this and continue welding. It can weld 15 different body styles that appear randomly on the assembly line.

CNC Machine Loading: A specially equipped robot loads 320 pound graphite electrode connecting pins into a CNC turning lathe, then into a CNC taper threading lathe followed by a CNC drill. Finally, the robot rotates the finished part and places it at the output station.

Parts Handling: The robot lifts car bodies off an assembly line, then rotates the bodies and hangs them on a 9-foot high conveyor.

ROBOT CHARACTERISTICS

Intelligent robots have six basic characteristics by which they may be classified. These are: structure, drive mechanism, end effector, controller, teaching system, and sensors.

Structure

The way the robot body is constructed can be one of two forms; polar or cylindrical.

Polar structure robot's
movements are limited to
two body movements and
the extension/extraction of
its arm. An angle and ra-
dial distance describe its
movements.

Polar structure

In the polar configuration, as shown in *Figure 9-4*, the body of the robot can pivot vertically or horizontally or both. The robot arm, which extends radially from the body, can be extended and retracted. The wrist may or may not be movable. The key is that the body itself pivots in up to two planes and the arm only moves radially.

**Figure 9-4.
Polar Structure for a
Robot**

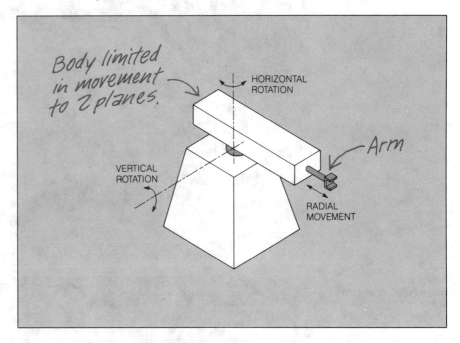

The robot shown in *Figure 9-5* is an example of a robot with a polar structure. This robot can rotate up to 200 degrees in the horizontal plane and the arm can extend out to 18 inches. It is microprocessor controlled and is used primarily for handling, feeding or mating of light parts. Large robots also can be of the polar configuration. The robot shown in *Figure 9-6* is one of these and is used extensively for material handling.

Cylindrical Structure

In the cylindrical structure shown in *Figure 9-7*, the basic body structure is a cylinder with an arm extending from the body. The body usually can rotate in the horizontal plane. The arm often can move in both the vertical and horizontal planes.

The intelligent robot shown in *Figure 9-8* has a cylindrical structure. The robot can lift up to 220 pounds and rotate through 240 degreees. The arm can extend over 112 inches vertically and 74 inches horizontally.

The type of structure usually dictates the type of coordinate system that is used to describe the movements of the robot. In the polar configuration, an angle and radial distance are specified. In the cylindrical configuration, X, Y and Z Cartesian coordinates are used to describe arm movement.

Cylindrical structure
robots usually move in the
horizontal plane and their
arms move vertically and/
or horizontally, as well as
radially. X, Y and Z coordinates describe its
movements.

Figure 9-5.
Polar Structured Robot
(Courtesy of Copperweld Robotics)

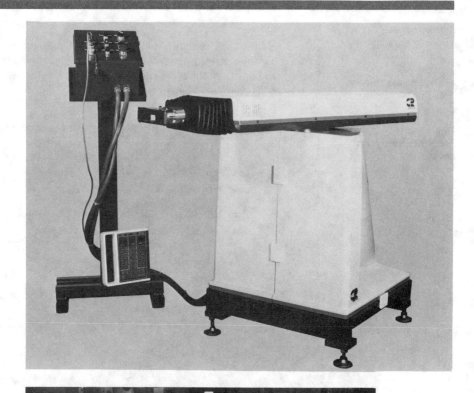

Figure 9-6.
Large Polar Structured Robot
(Courtesy of Unimation Inc., Danbury, CT)

**Figure 9-7.
Cylindrical Structure for
a Robot**

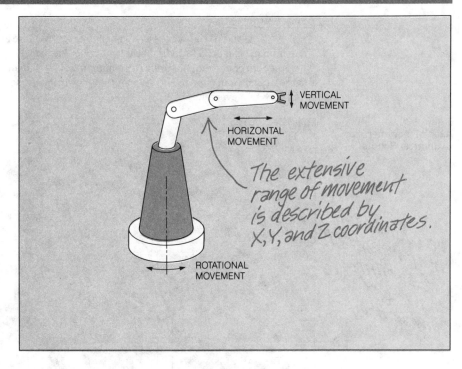

VERTICAL
MOVEMENT

HORIZONTAL
MOVEMENT

*The extensive
range of movement
is described by
X, Y, and Z coordinates.*

ROTATIONAL
MOVEMENT

**Figure 9-8.
Cylindrical Type Robot**
*(Courtesy of MTS Systems
Corporation)*

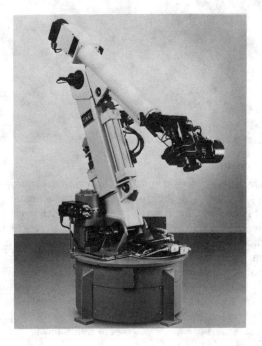

Axes of Movement

Most industrial robots are designed to try to move and do the same thing humans can do. The cylindrical type robot has a mechanical structure with freedom of movement similar to a human waist, shoulder, elbow, wrist and fingers as shown in *Figure 9-9*. As mentioned previously, some robots have up to six axes of movement.

Figure 9-9.
Axes of Movement for
Cylindrical Robots

The cylindrical robot movements are similiar to those of a human body.

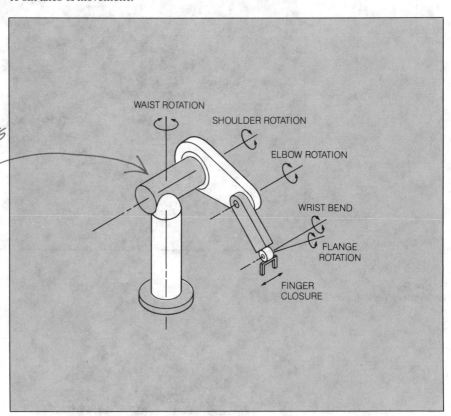

Drive Mechanism

In addition to three types of power sources for robots (electric, pneumatic or hydraulic), robot drives also can be servo controlled.

The power source that drives the manipulator (arm) of the robot usually is pneumatic, hydraulic, or electric. The pneumatic drive is the least expensive to purchase, but it is the most limited with respect to payload weight. Electric drives are the most expensive to purchase, but can handle heavy payloads and are the least expensive to operate. Hydraulic drives fall between pneumatic and electric with regard to cost and payload capability.

Large robots are almost always electrically driven and require significant electrical energy. The robot of *Figure 9-3*, for instance, uses 22 KVA at 230 or 460 VAC. Some pick-and-place robots such as the one shown in *Figure 9-1* use mechanical drives. The robot in *Figure 9-5* has a pneumatic drive mechanism. The one in *Figure 9-8* is hydraulically driven and requires 3 KVA at 460 VAC.

The drive mechanisms can be either servo or non-servo controlled. A servo controlled robot can be programmed to accelerate, decelerate or stop anywhere within its normal range of movement. The robots shown in *Figure 9-3* and *Figure 9-8* are servo controlled robots.

The non-servo controlled robot stops at fixed points on each axis. The robot shown in *Figure 9-5* has non-servo controlled drive mechanisms and has two fixed stops (that can be adjusted initially) on each axis of motion. This type of robot certainly possesses less flexibility of programming and application, but is simple to use and less expensive to purchase and operate. This brings up a significant factor when selecting a robot. To have a robot with many advanced features may be impressive, but if they are not necessary for the application and go unused once the robot is installed, then money has been wasted.

End Effector

The actual work the robot does is usually performed by its hand or, commonly called, an end effector.

The end effector is the robot "hand"; that is, the device that handles the payload. A wide variety of end effectors are available as standard off-the-shelf items and many others can be custom designed for a particular application. As shown in *Figure 9-10*, the end effector may be a mechanism for gripping a payload or it may be a tool (paint sprayer, welding gun, etc.).

**Figure 9-10.
Assortment of Robot
Grippers and Tools**

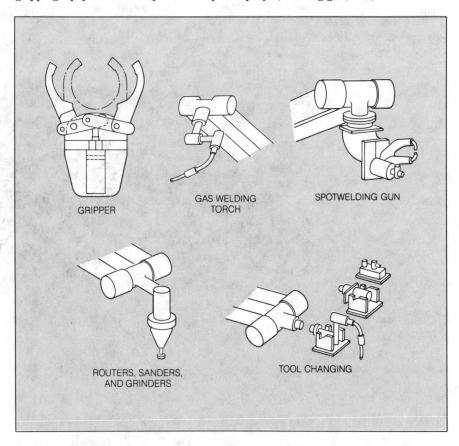

GRIPPER

GAS WELDING
TORCH

SPOTWELDING GUN

ROUTERS, SANDERS,
AND GRINDERS

TOOL CHANGING

Speed

The speed at which a robot's end effector moves often determines how productive it actually is.

Speed refers to the velocity of the tip or center of the end effector. If the end effector has to move over a long distance between points of action, as in pallet loading, then speed is the primary determinant of robot productivity. Other applications, such as spot welding, require only very small movement between points and speed has little effect on productivity. Most robots have speeds in the range of 30 to 50 inches/second, but a few are very fast, such as the Westinghouse series 6000, which can move up to 1,000 inches/second.

Controller

The heart of the intelligent robot is the controller. The controller may be a very simple (little intelligence) programmable controller such as the 5TI or it may be a sophisticated minicomputer such as the TI990/12 shown in *Figure 9-11*. The high-technology robots now coming into the marketplace use this type of minicomputer so that they are capable of very complex operation. In particular, robot vision systems are practical only because of minicomputer controllers.

**Figure 9-11.
TI 990/12 Minicomputer**

Intermediate-technology controllers are essentially CNC devices such as those discussed in Chapter 7. *Figure 9-12* shows a CNC driven robot with the controller located in the rack on the extreme right of the picture. Most computer-driven controllers are located remotely from the robot. This is due not only to size, but also to the hostile environment in which the robot often works (heat, dust, etc.).

Teaching System

Robots can be taught either by "walking through" the operation to be performed with a human teacher or by giving it instructions through computer software programs.

Basically, two teaching systems exist to teach the robot what it is expected to do. The first is the "walk-through" method where the robot arm is physically moved through the step-by-step set of operations it is to perform while the movements are recorded on magnetic tape or disk. This method is quick and requires no debugging of a program. However, the robot may need to be moved through the operations several times before achieving satisfactory operation from the tape playback. Also, large robots are difficult to handle and means for indicating things such as grip closure must be available.

The second teaching method is the software program. For simple robots using controllers such as the TI510, programming involves relay logic ladder diagrams. CNC systems in many cases (*Figure 9-12* for example) are based on the APT language. Sophisticated systems using minicomputers such as the TI990/12 rely on FORTRAN, PASCAL, PL/1 or English language type software.

Figure 9-12.
Robot Controller Located
Remote from Robot
(Courtesy of Westinghouse
Electric Corporation)

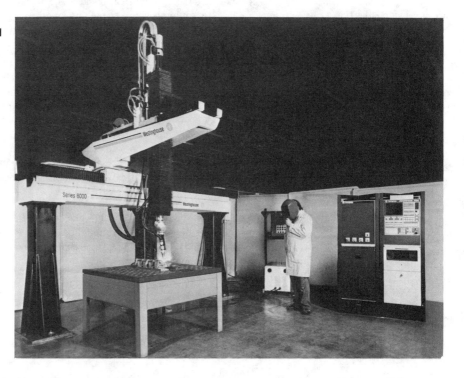

Program Languages

There are a great many different types of computer programs used in robots; some very sophisticated, some simple. It will be a long time before any universal robot language is in general use.

All major robot manufacturers have developed their own specialized program language for use with their robots. Most languages available are point-to-point where the user stores a series of points that guide the robot through a sequence of operations. Other languages are on the primitive motion level - one that allows simple branching, subroutine calls and sensing capabilities. A few languages exist at the structured program level which allows complex calculations such as coordinate transformations and may allow the use of vision systems. IBM is in the process of developing a task-oriented language with English-like instructions[1]; e.g., "pick up object 1" and "place on object 2."

Languages used to program robots are difficult to understand and apply without special training; therefore, specific programming examples for robots are not included in this book.

Some effort is underway to develop a universal robot language, but manufacturers usually resist the acceptance of a language that is much different from their own. Consequently, it will probably be several years before a common language is universally accepted. Thus, a primary consideration in purchasing a robot is the teaching system support offered by the manufacturer. Is it sufficiently sophisticated to handle the application? Is it easy to use?

Sensor

Vision sensing and touch are areas in robotics that still need significant development.

As robots are designed into applications, the need to sense the position of objects, to apply pressure to pick up an object, or to find one object among others becomes apparent. Particularly important are the senses of touch and sight. Pressure sensors mounted in the hand of the robot provide information feedback to adjust grip pressure to prevent dropping or crushing the object in the hand. Also, applying twisting force (torque) or sensing the distance between objects is necessary in some applications.

Vision systems are more complex and require a computer to process the image seen by the robot "eye". Image processing and analysis techniques change the visual image generated by a TV-type camera system to a form that can be used by the robot controller. The results are used to make decisions concerning the robot's next actions. The camera may be mounted on the robot itself or may be located remotely, depending upon the application.

Specifications

When selecting a robot for a particular application, a set of specifications based on the application should be developed before considering any specific robot. Some of the specifications that would be included in a specification list are shown in *Table 9-1.*

**Table 9-1.
Sample Specifications of
Commercially Available
Robots**

	Drive System	Control System	Path Control	Memory Capacity	Axes of Movement	Maximum Payload (lb.)	Speed In./Sec.	Repeatable Accuracy
CINCINNATI MILACRON T³	Hydraulic/Electric	Computer Servo	Continuous	700 pts.	6	100	50	±0.05
UNIMATION UNIMATE 2100B	Hydraulic/Electric	Computer Servo	P-to-P Continuous	2048 pts.	3 to 6	300		±0.08
SEIKO 200	Pneumatic	Mechanical	Stops	2/axis	4	1.65	30	±0.0004
WESTINGHOUSE 6000	Electric	Computer Servo	P-to-P Continuous	500 pts. 64K words	5	100	1000	±0.005
MTS 200A	Hydraulic	Computer Servo	P-to-P Continuous	2000 pts.	4 to 6	220	0-180° per second	±0.025
THERMWOOD SERIES 3	Hydraulic	Computer Servo	P-to-P	200 pts.	5	50	30	±0.06
UNIMATION PUMA		Computer Servo	Continuous	64K words	5 to 6	5	40	±0.004

WHEN ROBOTS ARE BENEFICIAL

There are many benefits that can be realized by adding robots to a manufacturing operation. The most obvious of these benefits are those found in improved quality, productivity and a reduction in human risk in adverse conditions.

Even though robots are not a universal solution to production problems, economic woes and foreign competition, they offer many benefits to a company. Some of these are summarized in the following paragraphs.

Increased Production Quality

The robot is not concerned about working conditions and does not get tired during the workshift. Its work is repeatable and predictable. The human's Monday and Friday syndrome of poor quality and production does not exist with robots. Production quality is uniform.

Increased Productivity

Robots do not take coffee breaks, can work with a consistent rhythm and rarely get sick. Because of this, use of robots almost always results in increased productivity.

Work in Unpleasant or Hazardous Environment

One of the major advantages of robots is their ability to withstand environmental conditions that are either unpleasant to humans or not allowed because of the requirements of the Occupational Safety and Health Administration (OSHA). Robots can withstand much higher levels of heat, dust, radioactivity, noise and noxious odors than humans without any degradation in quality or productivity.

Increased Image

Not to be overlooked is the public relations aspect of robot utilization. It can increase the positive image of a company in the eyes of the stockholders and customers. It indicates that the company is on the forefront of technology and may imply that the product can be expected to be technologically advanced in other areas.

Reduction in Training Costs

Most robots are easily trained and retrained to perform tasks. Training and retraining a human often requires more time and is more expensive.

Minor Facility Modification

Robots usually can be installed in existing plants with only a few changes required. Alteration of the assembly line to provide workspace for the robot may be necessary and larger wires for feeding power may be required, but completely new plant facilities are not necessary. The workspace required by a robot is called its "work envelope." *Figure 9-13* shows this information for a particular robot with a cylindrical structure and *Figure 9-14* shows the workspace requirement for a particular robot with a polar structure.

**Figure 9-13.
Workspace
Requirements for
Cincinnati Milacron T³**
*(Courtesy of Cincinnati
Milacron)*

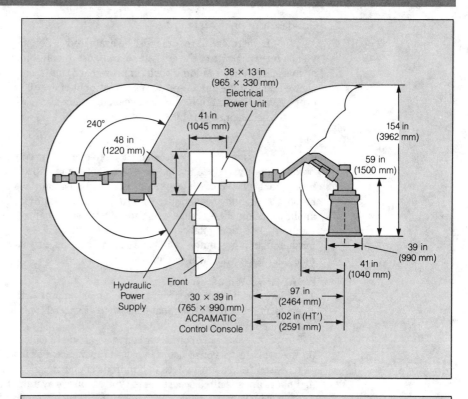

**Figure 9-14.
Workspace
Requirements for
Unimation 2000B**
*(Courtesy of Unimation Inc.,
Danbury, CT)*

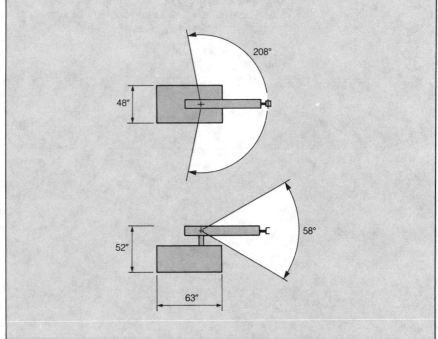

Good Return on Investment

Investment in robots must be evaluated to determine how fast they yield a return. Some of the analyses are almost unbelievable. For instance, one analysis involving a welding robot improved arc-time from 30% to 70% and showed an average annual return on investment of over 600% when evaluated over a five year period. Even a return on investment of 15% to 30% may indicate that the use of robots is justifiable.

Summary of Benefits

In summary, robots can be used to increase productivity, increase product quality, reduce training requirements, and improve return on investment when correctly applied. Likely applications are those that have harsh work conditions and jobs that require unusual physical effort. But these applications must be evaluated carefully to be sure that robots truly are required. The functional capability of robots will increase as new designs are marketed and they indeed will have a significant part to play in the manufacturing system of the future.

WHAT HAVE WE LEARNED?

1. An industrial robot is a programmable manipulator with a capability to perform tasks by moving material parts or special tools in a prescribed manner.
2. Most new robots are computer controlled and are called intelligent robots.
3. Most robots are constructed in polar or cylindrical form and may manipulate payloads from a few ounces to hundreds of pounds.
4. Robots are specified according to type of control system, memory capacity, axes of movement, payload range, speed, and repeatable accuracy.
5. The return on investment in robots can be very high.

Quiz for Chapter 9

1. Which of the following devices would be classified as a robot by the Robot Institute of America?
   a. Automobile.
   b. Pick-and-place manipulator.
   c. Artificial hand.
   d. All of the above.

2. An intelligent robot can:
   a. make decisions.
   b. handle only small payloads.
   c. speak.
   d. walk.

3. The physical structure of a robot can be of:
   a. angular or perpendicular form.
   b. perpendicular or polar form.
   c. cylindrical or polar form.
   d. any of the above.

4. The power source that drives the manipulator can be:
   a. electric.
   b. pneumatic.
   c. hydraulic.
   d. all of the above.

5. A servo controlled robot:
   a. stops only at fixed points on each axis.
   b. can accelerate.
   c. must be electrically driven.
   d. must be hydraulically driven.

6. Which of the methods below can be used to tell a robot what it is to do?
   a. Walk-through.
   b. Set mechanical stops.
   c. Software program.
   d. All of the above.
   e. a and c.

7. Programming languages currently available for robots:
   a. may be point-to-point languages.
   b. require training.
   c. may allow primitive motion.
   d. none of the above.
   e. all of the above.

8. Advantages of robots include:
   a. increased productivity.
   b. ability to withstand hazardous environments.
   c. good return on investment.
   d. better company image.
   e. all of the above.

9. Robots can increase productivity because:
   a. they do not get tired.
   b. they don't take a vacation.
   c. production quality is uniform.
   d. all of the above.

10. Intelligent robots:
    a. can adjust grip pressure.
    b. have vision capability.
    c. can sense distance between objects.
    d. all of the above.
    e. none of the above.

11. A pick-and-place manipulator is:
    a. an intelligent robot.
    b. the most flexible of all robots.
    c. capable of point-to-point operation.
    d. primarily used for large parts handling.

12. The accuracy of a pick-and-place robot can be as good as:
    a. ± 0.0005 inches
    b. ± 0.01 inches
    c. ± 0.005 inches
    d. ± 1%

13. The "walk-through" method of programming a robot involves:
    a. walking through your suggested procedure with your supervisor prior to programming.
    b. physically moving the robot through all the motions it is to repeat.
    c. moving the robot from one location to another.
    d. all of the above.

14. The difference between a high-
    technology and a medium-
    technology robot is the:
    a. range of motion of the robot.
    b. speed of movement.
    c. accuracy of the robot.
    d. complexity of the task it can
      perform.

15. The structure of a robot:
    a. determined its potential uses.
    b. defines its range of motion.
    c. defines the workspace needed for
      the robot.
    d. all of the above.

16. Robot motion:
    a. is the same for all robots.
    b. usually imitates human motion.
    c. is not dependent on the robot
      structure.
    d. none of the above.

17. The axis of movement of a robot may
    include:
    a. X-Y coordinate motion.
    b. wrist rotation.
    c. elbow rotation.
    d. all of the above.

18. The "end-effector" of a robot:
    a. is the robot "hand".
    b. can be an actual tool.
    c. may have a gripping action.
    d. all of the above.

19. Robots are specified by their:
    a. control system.
    b. payload.
    c. axes of movement.
    d. all of the above.

20. The workspace required for a robot:
    a. is less than that of a human
      performing the same task.
    b. depends upon the robot used.
    c. must be enclosed by a fence.
    d. is not dependent on the
      application.

Automation Example

ABOUT THIS CHAPTER

This chapter deals with the automation of an assembly line for manufacturing discrete parts. Automation of an assembly line is usually considered when the volume of a product is high enough and anticipated product life is long enough to justify spending the money required to develop the automation equipment. The successful use of automation not only reduces the assembly time, which can increase the product output per day, but also reduces the cost of manufacture of that product. A lower cost product, in turn, often increases the demand for the product because more people are willing to buy the product at the lower price. The semiconductor and electronic equipment industries are good examples of this effect.

This chapter places emphasis on the use of a microcomputer as a controller in the automation process. Other books are available for the reader who has an interest in the mechanical aspects of the automation task. This is not to say that the mechanical portion is a small portion of little concern, in fact, just the opposite is true. However, the microcomputer permits the mechanical portion of the design to be less complex and allows separation of the control function from the actuator itself. This means that the function of the actuator can be changed easily by changing software rather than requiring a redesign of the actuator mechanical control. This advantage often is not obvious when an automation system is first designed, but later it is appreciated.

AUTOMATION SYSTEM DEVELOPMENT PROCESS

A complete understanding of the process to identify what is to be controlled, its requirements, and how it is to be controlled is the most important step to assuring the success of the design of an automation system.

There usually is more than one method to solve most manufacturing problems. Deciding on the method that will be used is necessary, but the most important objective when considering the automation of a process is to ensure that the process is well understood. The biggest cause of failure in automation design is not the failure of the design, but rather the failure to fully anticipate and specify the system requirements before the design.

The best controller in the world is no good if it is incorrectly chosen and applied; therefore, it is important to properly specify what is to be controlled and how it is to be controlled before specifying the controller. The use of electronics in control systems has made it possible to control almost any operation easily and effectively. Now it is not so much necessary to determine if it is possible to control an operation of a process as it is necessary to determine that actual physical hardware can be chosen to do the control. Once this is determined, more than likely the operation or process can be controlled with an electronic controller.

System Requirements

To determine how large and complex the control system must be to adequately perform the desired control level, it is best to write a set of specifications for each process step.

The first step is to determine an overall system requirement for the control. For instance, if one stand-alone machine with its controller will perform the automation task, then there is no need for a complex communications network with a central controller. On the other hand, if the assembly process involves the use of many machines covering a large area of the plant, communication links to a central controller as shown in *Figure 10-1* might be a requirement. The central controller can be a minicomputer, like the Texas Instruments TM990 series, a microcomputer like the TM990/100 series shown in the example in this chapter, or a 5100 supervisory controller for a programmable controller system.

Figure 10-1.
System Block Diagram

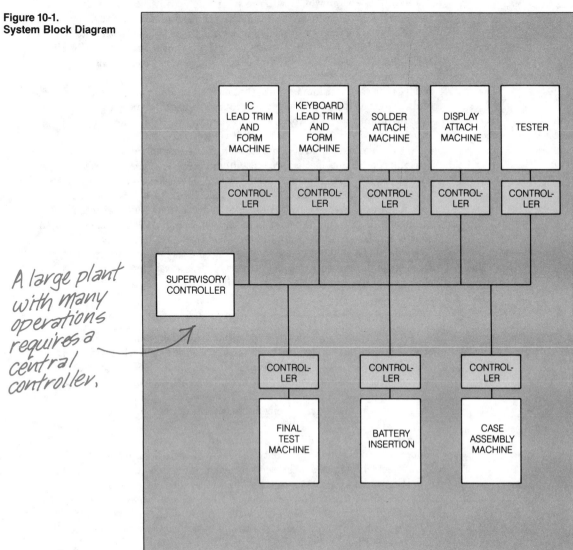

A large plant with many operations requires a central controller.

A great temptation to the designer is to require more "bells and whistles" (optional luxury features) on the system simply because they are available rather than actually necessary. However, the designer must understand that the difficulty of a design increases rapidly as the complexity increases, so it is good practice to keep it as simple as possible. This is especially true for items like communication links that are necessary, but not vital.

The best way to resolve what is required is to write a detailed list of the specifications. Cover each item in as much detail as possible, then as requirements become firm, the specifications can be shortened or reduced to eliminate unnecessary material. Some systems can be specified easier by a top-down approach; while for other systems, it is best to build the list upward from the individual parts to the overall system.

Control System Development

After the overall system requirements are determined, the individual machine control requirements can be defined. In the past, because systems required extensive communications links or data manipulations, microcomputers were required because a programmable controller was not adequate. However, as the capability of the PC has increased, this dividing line is less clear and other parameters have to be used to determine which controller to use. Of course, the capability to meet the system specifications is the first requirement; but cost, availability of service, and maintenance or exchange support are equally as important. Another important, but easy to overlook item, is software development support.

Software Development

The type of control required, how fast the system must operate and whether an existing operating system can be used are factors that influence the microprocessor that is selected and the software that supports it.

A typical controller software development flow is shown in *Figure 10-2*. First, the machine control in terms of operation, speed and I/O interface points must be determined. A flow diagram of the machine operation may be helpful in this task. Speed of operation can be important to the programming language that is chosen if a microcomputer controller is used. Usually, programmable controllers have their own language as has been discussed. If a high-level language is used for microcomputer programming, it usually does not execute as fast as assembly language programs; therefore, if speed is critical, assembly language programming may have to be the choice.

After the type of processor has been chosen and the programming language decided, usually another decision has to be made regarding the operating system software. If a large automation system is to be designed, usually it is best to use an operating system that has already been prepared for the particular microcomputer being used as a controller. For small systems, the operating system code may be written directly into the total program for the system. However, this could greatly increase the reprogramming effort when the system is changed. The reprogramming effort is less when a prepared operating system is used because it already is programmed to handle many input/output variations.

**Figure 10-2.
Software Development
Flow Diagram**

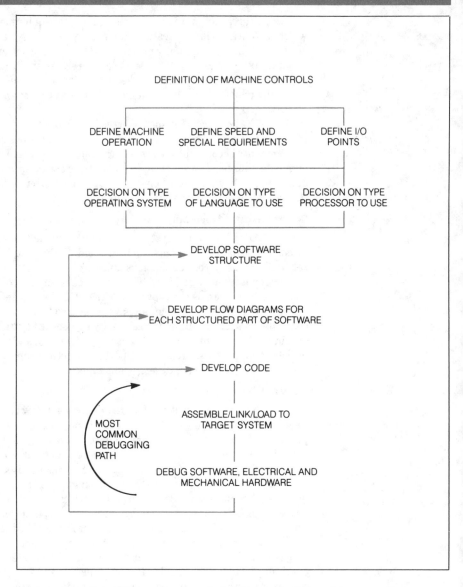

DEFINITION OF MACHINE CONTROLS

| DEFINE MACHINE OPERATION | DEFINE SPEED AND SPECIAL REQUIREMENTS | DEFINE I/O POINTS |

| DECISION ON TYPE OPERATING SYSTEM | DECISION ON TYPE OF LANGUAGE TO USE | DECISION ON TYPE PROCESSOR TO USE |

DEVELOP SOFTWARE STRUCTURE

DEVELOP FLOW DIAGRAMS FOR EACH STRUCTURED PART OF SOFTWARE

DEVELOP CODE

MOST COMMON DEBUGGING PATH

ASSEMBLE/LINK/LOAD TO TARGET SYSTEM

DEBUG SOFTWARE, ELECTRICAL AND MECHANICAL HARDWARE

Once the software structure is decided upon, the machine operations are described with flow diagrams. Several levels of flow diagrams may be required with the lowest level being as close as possible to the instruction level required in the program.

From the flow diagram, actual program code is written. Many designers have the idea that the first thing they should do is write program code, but notice that *Figure 10-2* shows the coding step far down in the development cycle.

It is a rare event when a program runs exactly the way the designer wants the first time; therefore testing, debugging and modification usually are necessary. Thus most commonly, several passes may be required through the loop of the last three steps of *Figure 10-2*. However, if problems are severe, some redevelopment of the software structure and/or the flow diagrams may be required.

In large systems, the program actually may be developed on a separate development system that simulates the actual system. In this case, the completed program must be loaded into the actual microprocessor system RAM or installed permanently in ROM.

EXAMPLE OF AN AUTOMATED ASSEMBLY LINE

Introduction

One of the success stories in the electronics field is the development of the low-cost handheld calculator. When this product was first introduced to the marketplace, it was expensive and only a few people were willing to pay the price. Today, very sophisticated units with much more functional capability can be purchased for well under $15 and almost everyone owns at least one handheld calculator.

An automated final assembly process has contributed significantly to lowering the cost and increasing the sales volume of handheld calculators.

How did this come about? Much of the credit must be given to the integrated circuit designers and process engineers who were able to place more electronic functions on the semiconductor chip itself, which drastically reduced the number of discrete parts that had to be manufactured separately and assembled. Another important reason, however, was the development of automated production lines which allowed the product to be built faster, in higher volume, and at a lower cost than would otherwise be possible. Basically, the handheld calculator assembly has progressed from completely manual to almost totally automated. The automation includes even the testing and final packaging of the unit. Although this example will be concerned with the final assembly and test of the calculator, a similar description could be given for the manufacture of the individual components which make up the final assembly.

Assembly Flow Diagram

When analyzing the assembly flow diagram, keep an eye out for steps that are time-consuming, repetitive tasks. These are the most successful and productive steps to automate.

The assembly process through packaging shown in the diagram in *Figure 10-3* has a total of 14 steps. This includes four preassembly steps that must be performed on the keypad and the integrated circuit. In between each of these steps is a material transfer between stations, but this is not included in the 14 steps. In an automated system, the material transfer step may be just as difficult to perform as the process itself; in fact, this probably represents the biggest obstacle to automating an assembly line.

Automation systems are successful when they eliminate the more repetitive time-consuming tasks in the assembly process not only to reduce the amount of labor, but also to eliminate the tedious jobs that nobody wants. The remaining manual jobs then are more enjoyable.

**Figure 10-3.
Assembly Task Flow
Diagram**

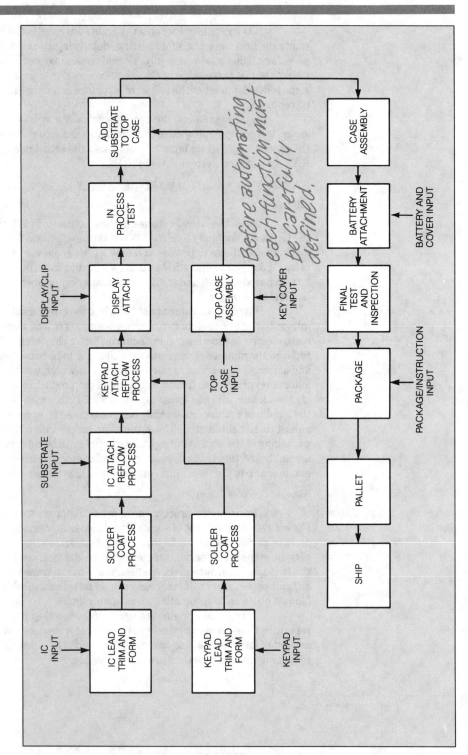

SYSTEM REQUIREMENTS

The first step in analyzing if an automated system is practical and worthwhile economically is to determine the time it takes to manually assemble a calculator.

In this example, the materials and parts were already specified and designed, so the requirement was to automate the assembly so that the calculator could be built to reduce the assembly time by a factor of 50.

Manual Assembly

The time taken to perform the operations manually was:

Operation	Time (seconds)
1. Trim and form integrated circuit	14
2. Trim and form keypad.	14
3. Solder coat integrated circuit leads.	6
4. Solder coat keypad leads.	6
5. Attach integrated circuit to substrate.	16
6. Attach keypad to substrate.	24
7. Attach display to substrate.	32
8. In-process functional test.	60
9. Assemble keypad to top half of case.	10
10. Assemble substrate with components into top case.	15
11. Place back of case onto top and glue.	12
12. Add batteries and battery cover.	12
13. Perform final functional test and inspection.	75
14. Package.	12
Total	308

Thus, the total manual assembly time for these steps was 5 minutes and 8 seconds per unit, or production at the rate of 11.7 units per hour per operator. The assembly cost per unit at this production rate was determined by summing the labor and overhead cost, the costs for material handling, inspection and control, and the line inefficiency caused by unequal loading at each station. The manual assembly cost was used to compare to the assembly and capital equipment cost of the automated approach to give an indication of the savings from automation. Although there were other savings, such as reduced management overhead, reduced floor space, etc. which could be considered in determining the overall cost savings involved; there also were repair, maintenance, and spare parts costs for the automation system which offset these savings.

Automated Assembly

The goals set for the automated assembly were:

Item	Goal
1. System cycle time:	6 seconds/unit
2. Mean Time Between Failures (MTBF):	12 hours
3. Mean Time To Repair (MTTR):	1 hour
4. Line availability:	90%
5. Preventive Maintenance Schedule:	once per day
6. Machine Life:	5 years
7. Operator Service:	Reload parts once per hour.
8. Flexibility:	Manufacture all anticipated products for 3 years.
9. Central Control:	Serial interface to all machines from central console.

The 37.5% cost savings that is realized by automating the final assembly line is enough cost reduction to justify developing the automated system. More extensive and more accurate process monitoring and reporting are additional benefits of the automated system.

By determining the number of units that would be produced in the anticipated 3-year product life using two 8-hour shifts per day for five days a week and 50 weeks per year, it was found that the automated assembly cost was 37.5% less than the manual assembly cost. The labor and overhead cost, design and development cost of the machine and capital equipment cost of the machine were included in the automated assembly cost. This was enough cost reduction to justify development of the machine.

In addition to the product assembly, a by-product of the automated system is the ability to monitor and report the status of both the machine itself and the amount of product assembled. Reports on the total assembled parts output, the reject rate at the intermediate and final test stations, the actual run rate of each machine, the inventory status, etc. will help the line supervisor determine if component problems exist or if more product output is possible.

Automation Machine Structure

Part of the task of the mechanical designer is to determine the best way to handle the individual assembly tasks. Decisions must be made on how to partition the steps, whether more than one machine is necessary, and which, if any, of the steps could be handled more effectively as a manual task. Consideration also must be given to whether the assembly process can be handled better in one continuous process machine or in steps where there are logical breaks.

The batch method was chosen for this system so that material flow could be balanced with additional stations at particular points and manual assembly could be substituted easily as a back-up.

In this application, the decision was made to separate the machines by functional task. A continuous process machine would result in total line shutdown if any one portion of it were to fail, but a batch concept permits manual assembly to be substituted for the failed portion of the process if one machine malfunctions while the rest of the line continues in the automated mode. This allows a much better backup effort with less standby labor and with less complex and less expensive machines. Since the continuous process would not improve the final product in this case, the batch method was chosen.

One or more separate machines is used for each process type; namely, lead trim and form, solder reflow, component attach, case assembly, battery insertion, test, and packaging. Testing requires more than one station because testing takes longer than the other process steps. This is true even when the testing is automated because the calculator cannot be tested faster than it can operate. The number of machines for each process were chosen so that no one machine has to be operated significantly longer than any other machine in order to maintain the required pace of the entire system. Let's look at the design process in a little more detail.

SPECIFICATIONS

All details of equipment operation, interface and performance must be specified to properly implement the system.

In order to implement the system, all the input and output parameters to the system including material flow, operator interface, and system interface must be specified. The performance of each machine including the speed, operations to be performed, utilities required, etc., must be determined. For example, material and machine tolerances must be specified so that the material can be handled properly.

Since different machine types are used in this product assembly, individual machine specifications as well as an overall system specification are needed. As an example of individual machine specifications, let's consider the solder reflow machine which is used for making electrical connections.

Solder Reflow Process

The solder reflow process itself is as follows: The parts that are to be joined are dipped in a hot liquid solder bath which coats the metal terminals with solder. Flux is applied to the two parts that are to be joined together to clean the surfaces. Next, heat is applied to the terminals to reheat the solder while maintaining the two parts in mechanical contact. The solder "reflows" between the metal terminals and the heat is removed. The joint is allowed to cool while maintaining the two parts in contact so that a new airtight and mechanically-strong bond forms. The solder reflow machine uses this process to join the integrated circuit and the keypad to the substrate.

Solder Reflow Machine Specifications

1. Cycle Time: 6 seconds or less
2. Solder Process Cycle: Apply solder flux, align parts, apply pressure, apply heat, cool under pressure, release part.
3. Solder Process Control: Operator inputs for heat time and cool time. All other parameters are defined by control engineer.
4. Alignment Accuracy: Part to be aligned on the substrate within 0.005 inch, adjustable at setup time.
5. Component Inputs: Substrate to be hand loaded, no special cannister required. Machine to detect if substrate is loaded incorrectly, and must handle two sizes of substrate. IC are supplied in tubes of approximately 20 devices each, all oriented correctly. Keypads are in metal cannisters of approximately 100 pieces. Cannister dimensions are defined to handle 5 types and prevent wrong orientation. Flux is required at two locations on substrate.
6. Component Output: No special handling is required. Next step will be hand loaded.
7. Machine Operation: Machine to do all material transport automatically and provide for diagnostic troubleshooting. Must be able to run the machine with and without material as required for set up and diagnostics.
8. Problem Recovery: Machine will self-diagnose all mechanical malfunctions and display problem to operator. Step-by-step mode and individual station control will be provided for repair and maintenance personnel.

SOLDER REFLOW MACHINE IMPLEMENTATION

A microprocessor digital system is used as the controller for the solder reflow machine because of its quick cycle time, ease in testing, remote communication capability, and mechanical design.

A digital microprocessor system was chosen for this application to allow simpler mechanical design, faster machine cycle time, easier testing and easier remote communications. The solder reflow machine will be used to illustrate microprocessor control of a machine. This machine is one of six machines that was developed for the assembly process. The other five are similar to this machine and all six use the *same* microprocessor as their controller.

CONTROLLER HARDWARE

There are a number of methods for implementing a digital controller. The most effective method is to use standard off-the-shelf board components so that very little detail design is needed. There are several advantages to this approach. Time and money are saved by using what has already been designed, built and proven and the design engineer is basically a system integrator to put the various components together to form the desired system. With this approach, the end user can select the particular items needed for the particular job. Standard microcomputer systems exist with a range of capabilities provided by various I/O boards, standard RS232 interfaces and memory expansion. The components are designed to work with each other so that

By using a commercially available TM990/102 microcomputer board that is a fully assembled tested unit, the designer can concentrate on the process rather than on designing and assembling the controller.

interfacing the different parts of the control system is not a problem. Most of the control system documentation is provided and this also saves time and money.

The entire control process of the solder reflow machine is handled by the single-board microcomputer (TM990/102) shown in *Figure 10-4*. This microcomputer uses the TMS9900 16-bit microprocessor as its CPU. Other features of this microcomputer are; extended addressing to one megabyte, EPROM expandable to 16K bytes, up to 128K bytes of on-board dynamic RAM, 16 prioritized vectored interrupts, and an RS232 port.

**Figure 10-4.
TM990/102
Microcomputer Board**

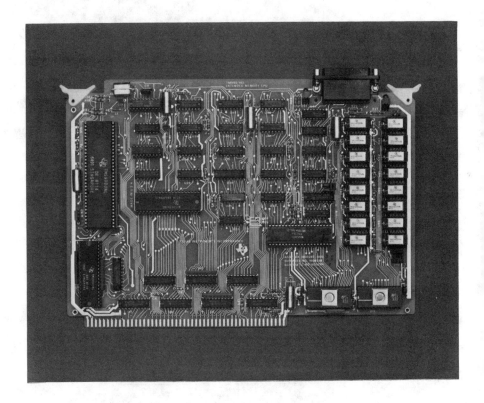

For this application, an RS232 communications board, three standard peripheral control boards, and three each of a custom designed stepping motor control board are used with the computer for I/O and communications. The physical arrangement of the control system is shown in *Figure 10-5*. *Figure 10-6* shows a block diagram of the same hardware. Power supplies, cables and I/O devices make up the remainder of the system. All of the I/O except for stepping motor control could be handled by standard peripheral boards offered with the microcomputer boards. The stepping motor control board had to be designed specifically for this application using a prototyping board available in the board family.

**Figure 10-5.
Control System Card
Cage**

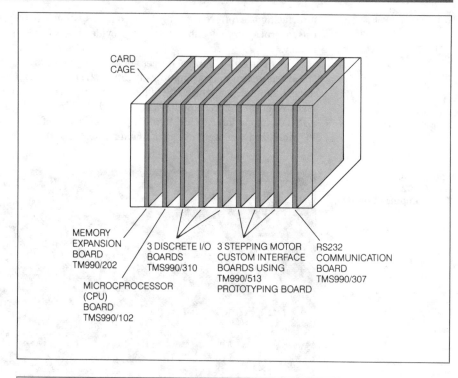

CARD
CAGE

MEMORY
EXPANSION
BOARD
TM990/202

3 DISCRETE I/O
BOARDS
TMS990/310

3 STEPPING MOTOR
CUSTOM INTERFACE
BOARDS USING
TM990/513
PROTOTYPING BOARD

RS232
COMMUNICATION
BOARD
TMS990/307

MICROCPROCESSOR
(CPU)
BOARD
TMS990/102

**Figure 10-6.
Control System Block
Diagram**

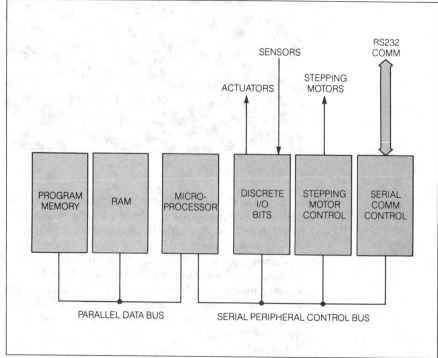

RS232
COMM

SENSORS

ACTUATORS

STEPPING
MOTORS

PROGRAM
MEMORY

RAM

MICRO-
PROCESSOR

DISCRETE
I/O
BITS

STEPPING
MOTOR
CONTROL

SERIAL
COMM
CONTROL

PARALLEL DATA BUS

SERIAL PERIPHERAL CONTROL BUS

Except for the stepping motors and the heating elements, which have their own control elements, the TM990/102 provides all timing and control functions.

Most of the timing and control for the system is accomplished by the microcomputer under program control. However, the stepping motors are sensitive to time variations and require critical timing for proper operation. It is difficult to control the motor timing using software control since the microcomputer execution times vary depending on which set of program instructions are followed. Because of this, timing is controlled by hardware logic on the stepping motor control boards. The output switches on the board are discrete transistors capable of supplying several amperes in order to drive the motors at the speeds required to meet the overall timing specification.

The two heater elements for the solder reflow operations are individually controlled with PI control to maintain a precise temperature. Each heater controller provides information on the temperature to the microcomputer so the microcomputer can maintain overall system control.

CONTROLLER SOFTWARE

The software is developed and implemented at several different levels in separate modular-like packages.

The software development for this controller is done in a structured manner as shown in *Figure 10-7*. Several levels of software are used to control the machine and each level is developed so that it meets well-defined software interface requirements. This allows several different programmers to develop different levels of the program independently of one another and have it operate successfully when it's all put together. When the software is developed into modules like this, usually the software can be used over again in other applications. Some of these modules are discussed in the following paragraphs.

**Figure 10-7.
Overall Software
Structure**

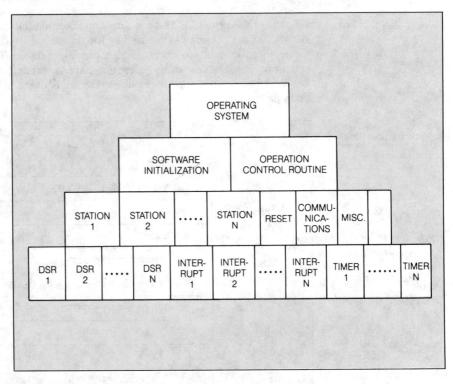

Operating System

The operating system keeps all system activities working in the correct sequence, passes information between different software routines and keeps track of basic operating functions.

An operating system is either developed or else one is used that is available from the microprocessor manufacturer. An operating system is a software program that keeps everything in the system operating together and in the correct order. It passes information between different software routines as the machine is operating and keeps track of basic functions like generating a timer delay or checking for errors. Thus, it allows short software subroutines to be developed and then put together into a complete control program that can execute very complex tasks.

Much has been written on operating systems, how they are developed and how they are used. It is not the purpose here to go into great detail (other books are available for that), but to indicate that an operating system exists, that it is very important to the system operation, and that it is used by the technician and engineer to develop the system software.

Function Control Block

The function control block is a block of predetermined information that helps the communication between the operating system and the subroutines.

In order for the operating system to properly handle the subroutines as they are required in the overall program, each subroutine will communicate with the operating system by having a block of information called a function control block (FCB) associated with it. This block of information, as shown in *Figure 10-8*, has a particular arrangement that has been decided on beforehand. Byte 1 contains the name of the subroutine. Byte 2 is a spare location in case it is needed at some future time. Byte 3 contains data on the subroutine or system status that the operating system checks as the subroutine is executed. Byte 4 contains a number that tells the operating system how many variables are used by the subroutine. Byte 5 points to the memory location address where the first variable is stored. This is called a pointer because the operating system may have to search through the memory locations from the initial pointer value to the end of the variables locations to find a particular variable.

**Figure 10-8.
Function Control Block**

The byte order was determined before programming.

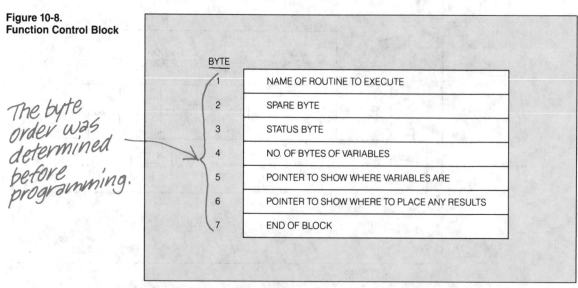

BYTE	
1	NAME OF ROUTINE TO EXECUTE
2	SPARE BYTE
3	STATUS BYTE
4	NO. OF BYTES OF VARIABLES
5	POINTER TO SHOW WHERE VARIABLES ARE
6	POINTER TO SHOW WHERE TO PLACE ANY RESULTS
7	END OF BLOCK

As the subroutine is run as part of the main program, results of the subroutine must be stored in a specific location so another subroutine will know where to find these results. Byte 6 points to the first memory location where these data are stored. Byte 7 marks the end of the FCB.

Operation Control Routine

The operation control routine (OCR) ensures that the system, particularly the CPU, is operating as efficiently as possible. It decides the specific tasks and the sequence that the CPU performs them.

The routine that supervises all of the programs that control the system hardware to do applications tasks is called the operations control routine (OCR). It decides which specific application task to execute at any given time. The OCR is constantly switching from one task to another so that the CPU is kept busy and efficiently utilized. Although the CPU can process only one instruction at a time, it is very fast. However, the machine hardware usually takes a relatively long time to react to a controller command.

For instance, to run the transport shuttle, which will be explained later, requires the setting of only two bits. The controller can set the two bits in less than a millisecond. After the bits are set, it takes almost half a second for the transport shuttle to get to the end of its travel. During this time, the microcomputer typically can process over 5,000 instructions. Thus, if the software is properly structured, the controller can perform many other operations while the shuttle is moving instead of just idly waiting. In between the processing of other instructions, it can periodically check on the transport to determine when the end of travel is reached. This ability of the controller to perform several tasks at the same time can be compared to a juggler who keeps all the balls moving in step so that only one at a time requires action by the juggler.

Flags

A flag is a bit in a register or memory location that indicates one condition if it is set (set to a 1) or another condition if it is clear (reset to a 0). In other words, it's like raising or lowering a signal flag. A program subroutine can look at a particular flag to decide what it should do next.

OCR Operation

Flags alert the OCR that a specific task is to be executed if the flag is set to a 1, or not executed if the flag is set to a 0. When a flag is set to a 0 the OCR skips that step.

The operation of the OCR is illustrated in *Figure 10-9a*. The task flag pointer holds the number of the next task to be run. It is stepped by one each time the OCR executes. The OCR checks the run flag of the task currently pointed to. If the run flag is set, the task is executed; if the run flag is not set, the task is skipped. Let's assume that the task flag pointer is set to task 1 and that the task 1 run flag is set so that task 1 will be run next.

The OCR checks the task pointer to obtain the memory address that holds the task 1 program pointer. The OCR loads the contents of that address into the CPU program counter. The CPU then starts executing the task 1 program. Let's now assume an interrupt occurs so that task 1 is stopped before completion. The address of the next task 1 instruction is saved in the task 1 program pointer location so when the task flag pointer again allows task 1 to run, execution of task 1 will begin wherever it was at the time of interrupt.

Figure 10-9.
OCR Example

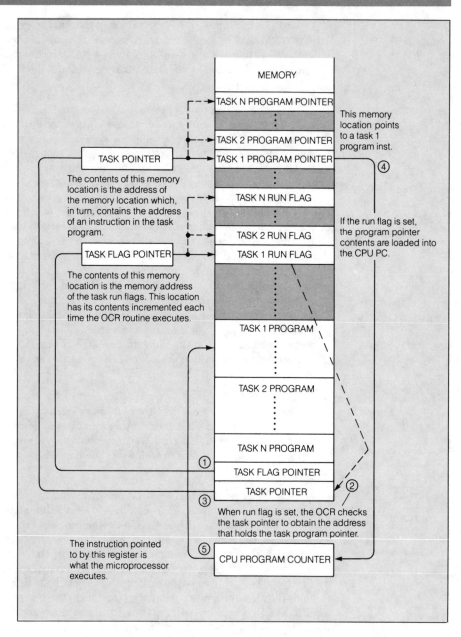

MEMORY

TASK N PROGRAM POINTER

TASK 2 PROGRAM POINTER

This memory
location points
to a task 1
program inst.

TASK 1 PROGRAM POINTER

TASK POINTER

④

The contents of this memory
location is the address of
the memory location which,
in turn, contains the address
of an instruction in the task
program.

TASK N RUN FLAG

TASK 2 RUN FLAG

TASK FLAG POINTER

TASK 1 RUN FLAG

If the run flag is set,
the program pointer
contents are loaded into
the CPU PC.

The contents of this memory
location is the memory address
of the task run flags. This location
has its contents incremented each
time the OCR routine executes.

TASK 1 PROGRAM

TASK 2 PROGRAM

TASK N PROGRAM

①   TASK FLAG POINTER

TASK POINTER   ②

③

When run flag is set, the OCR checks
the task pointer to obtain the address
that holds the task program pointer.

The instruction pointed
to by this register is
what the microprocessor
executes.

⑤   CPU PROGRAM COUNTER

Control is returned to the OCR which looks at the task flag pointer.
Since it was incremented by one, it now checks the task 2 run flag and finds it
set. The task 2 program pointer contents are loaded into the program counter
and the CPU begins executing task 2. Let's assume task 2 runs to completion
so the task 2 run flag is cleared and the task 2 program pointer is loaded with
its beginning address to prepare for its next execution.

Control is returned to the OCR and the sequence is repeated for each task until task N has been run or bypassed. Now the task flag pointer again points to task 1. Its run flag is still set since it wasn't completed, so the contents of the task 1 program pointer are loaded into the program counter and the CPU begins executing task 1 at the place it left when the interrupt occurred. Thus, the task programs are run in an interleaved fashion with each one running until something happens to stop it, typically an I/O operation, then the next is run, etc.

Device Service Routine

The device service routine (DSR) provides the timing signals and logic functions to interface the peripheral devices to the system. If the peripheral device changes, only the DSR need change, not the whole system.

Another routine used by the operating system for different input/output peripherals is called a device service routine (DSR). Manufacturers of microcomputers have standard DSRs available to handle many of the standard peripheral products such as a CRT or line printer. For automation systems I/O that is unique for a device such as a stepping motor, the user usually is required to develop a DSR which will handle operation of the device.

The DSR provides the proper timing and logic to the peripheral hardware. For instance, a DSR might be written to handle the logic signals from a keyboard to place the code for the particular key depressed in a buffer memory location specified by the user. The application program only needs to read the buffer to obtain the keyboard input. It doesn't need to know how to decode the keyboard or even to know where the input came from. If a decision is made to provide this input from a punched card reader rather than the keyboard, the application program need not be changed; instead only the DSR is changed to read the input from the card reader, and place the input in the same buffer. Thus, the same DSR might be used by many different application tasks.

A typical DSR is shown in *Figure 10-10*. It is used in the solder reflow machine to detect which keys are pressed on the keypad used by the operator to enter control parameters or ask for the machine status. The routine is activated by a FCB and results of the routine are placed in memory locations specified by the FCB. The instructions are 9900 family system instructions for the 9900 family microprocessor used in the TM990/102 microcomputer.

Application Programs

Individual application programs, which have been called tasks, handle the unique logic required for a specific portion of the machine. If the machine is simple enough, the task programs may be the only software written. In this example, a task is written for each functional block and each one runs only when called by the OCR. When execution of the task is completed, a status bit (flag) is set to notify the OCR. Usually other status bits indicate if operation was normal or if an error occurred.

**Figure 10-10.
DSR Example**

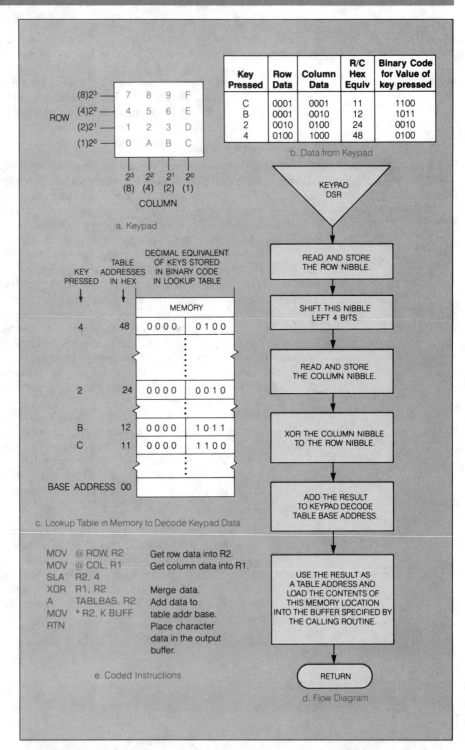

Key Pressed	Row Data	Column Data	R/C Hex Equiv	Binary Code for Value of key pressed
C	0001	0001	11	1100
B	0001	0010	12	1011
2	0010	0100	24	0010
4	0100	1000	48	0100

b. Data from Keypad

a. Keypad

c. Lookup Table in Memory to Decode Keypad Data

KEYPAD DSR

READ AND STORE
THE ROW NIBBLE.

SHIFT THIS NIBBLE
LEFT 4 BITS.

READ AND STORE
THE COLUMN NIBBLE.

XOR THE COLUMN NIBBLE
TO THE ROW NIBBLE.

ADD THE RESULT
TO KEYPAD DECODE
TABLE BASE ADDRESS.

USE THE RESULT AS
A TABLE ADDRESS AND
LOAD THE CONTENTS OF
THIS MEMORY LOCATION
INTO THE BUFFER SPECIFIED BY
THE CALLING ROUTINE.

RETURN

d. Flow Diagram

```
MOV  @ ROW, R2      Get row data into R2.
MOV  @ COL, R1      Get column data into R1.
SLA  R2, 4
XOR  R1, R2         Merge data.
A    TABLBAS, R2    Add data to
MOV  * R2, K BUFF   table addr base.
RTN                 Place character
                    data in the output
                    buffer.
```

e. Coded Instructions

THE TRANSPORT MECHANISM

The functions performed on this machine are:
1. Load a substrate.
2. Attach the integrated circuit to the substrate (two stations).
3. Attach the keyboard to the substrate (two stations).
4. Offload the assembly to the next assembly process.
5. Transport the substrate through the process steps and handle the component inputs.

The transport mechanism loads a substrate on a boat, transports the boat on a front track until assembled, unloads the boat, and transports the empty boat on a back track to the beginning position.

A shuttle mechanism was chosen for the transport mechanism to provide multiple stations on the machine while providing easy access to the components and allowing a compact size machine. The transport mechanism is a subsystem of the solder reflow machine, but the software is designed so that the control of the transport can be handled separately from the remainder of the machine. A top view outline drawing of the transport and the I/O points required to control it are shown in *Figure 10-11*.

At the load station, a cross-slide shuttle moves an empty metal transport "boat" from the back track to the load position on the front track. The boat is loaded with a substrate at the load station and locating pins in the boat ensure proper alignment of the substrate. The main shuttle operates a pawl mechanism which moves the loaded boats along the front track from the load station through the work stations to the unload station. After the completed subassembly is removed from the boat at the unload station, another cross-slide shuttle moves the empty boat to the back track. The main shuttle then returns the empty boats along the back track to the load station end in a circulating fashion. Sensors are provided at the four corners of the transport to ensure a mechanical jam does not occur in case of a malfunction. Position sensors at each shuttle cam sense the home position and the out (extended) position of each shuttle mechanism.

Software

Figure 10-12 shows the flow diagram and instructions necessary to program a cross-slide shuttle. The program for the other cross-slide shuttle and for the main shuttle are similar.

Error Handling

Both timer and sensor activated controls monitor flags and provide error alerts in case the transport mechanism does not operate properly.

The mechanical error routines monitor flags and timers to stop the machine if a malfunction occurs. For the transport to run safely, all of the assembly stations must be clear of the mechanism and in their home position. Each station has a flag which is either set to 1 or cleared to 0 to indicate whether it is running or at the home position. The transport control software checks all of these flags for the clear (CLR) state to ensure that all stations are at their home position before commanding any of the motors to run. Once any of the shuttle motors is started, normal operation will cause the motor to run until its position sensor, which indicates a cycle has been completed, is detected. If the shuttle mechanism jams for some reason so that the position sensor is not detected, a timer routine turns off the motor.

**Figure 10-11.
Boat Transport System**

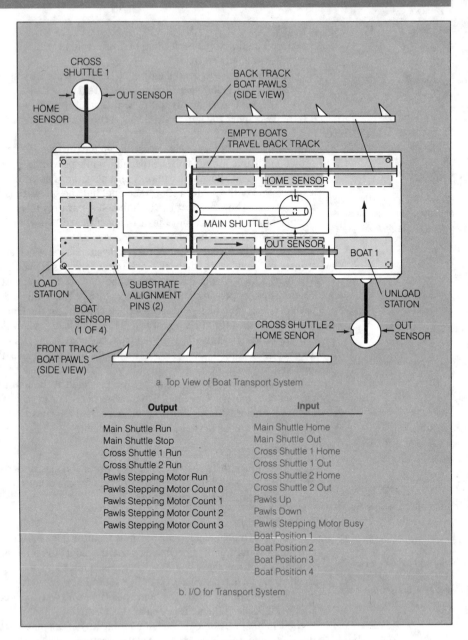

a. Top View of Boat Transport System

Output	Input
Main Shuttle Run	Main Shuttle Home
Main Shuttle Stop	Main Shuttle Out
Cross Shuttle 1 Run	Cross Shuttle 1 Home
Cross Shuttle 2 Run	Cross Shuttle 1 Out
Pawls Stepping Motor Run	Cross Shuttle 2 Home
Pawls Stepping Motor Count 0	Cross Shuttle 2 Out
Pawls Stepping Motor Count 1	Pawls Up
Pawls Stepping Motor Count 2	Pawls Down
Pawls Stepping Motor Count 3	Pawls Stepping Motor Busy
	Boat Position 1
	Boat Position 2
	Boat Position 3
	Boat Position 4

b. I/O for Transport System

The software timer routine, shown in flow diagram form in *Figure 10-13*, is started each time an operation is started. Enough time is allowed for normal completion of the operation; therefore, if the timer times out and the operation is not complete as indicated by the absence of a particular sensor signal, a timeout error is generated which causes the operation to be shutoff and the operator notified.

Figure 10-12.
Cross Shuttle 1 Routine

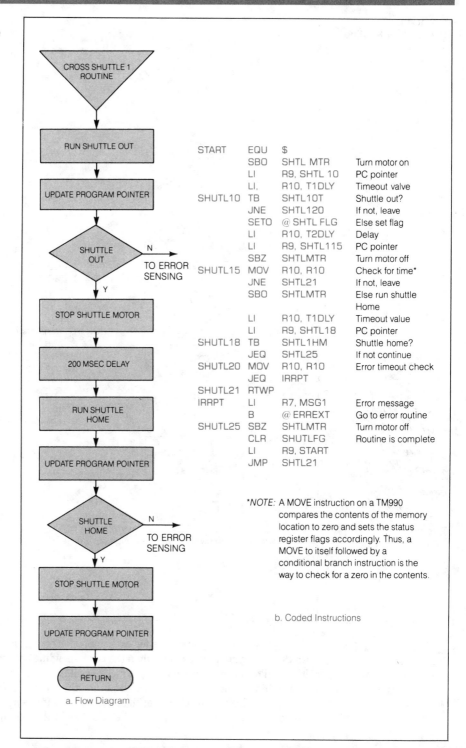

START	EQU	$	
	SBO	SHTL MTR	Turn motor on
	LI	R9, SHTL 10	PC pointer
	LI,	R10, T1DLY	Timeout valve
SHUTL10	TB	SHTL10T	Shuttle out?
	JNE	SHTL120	If not, leave
	SETO	@ SHTL FLG	Else set flag
	LI	R10, T2DLY	Delay
	LI	R9, SHTL115	PC pointer
	SBZ	SHTLMTR	Turn motor off
SHUTL15	MOV	R10, R10	Check for time*
	JNE	SHTL21	If not, leave
	SBO	SHTLMTR	Else run shuttle Home
	LI	R10, T1DLY	Timeout value
	LI	R9, SHTL18	PC pointer
SHUTL18	TB	SHTL1HM	Shuttle home?
	JEQ	SHTL25	If not continue
SHUTL20	MOV	R10, R10	Error timeout check
	JEQ	IRRPT	
SHUTL21	RTWP		
IRRPT	LI	R7, MSG1	Error message
	B	@ ERREXT	Go to error routine
SHUTL25	SBZ	SHTLMTR	Turn motor off
	CLR	SHUTLFG	Routine is complete
	LI	R9, START	
	JMP	SHTL21	

NOTE: A MOVE instruction on a TM990 compares the contents of the memory location to zero and sets the status register flags accordingly. Thus, a MOVE to itself followed by a conditional branch instruction is the way to check for a zero in the contents.

b. Coded Instructions

a. Flow Diagram

Figure 10-13.
Software Timer Routine

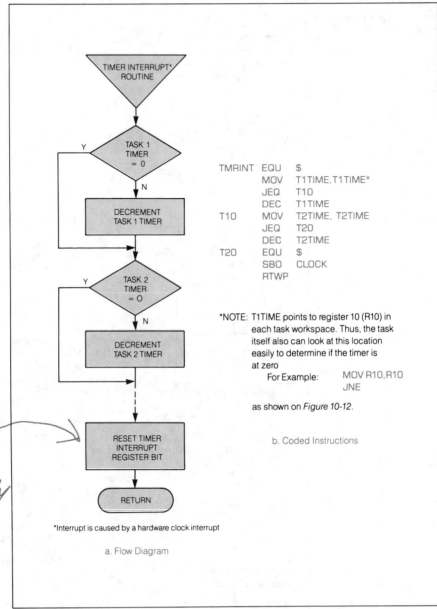

TMRINT EQU $
       MOV T1TIME,T1TIME*
       JEQ T10
       DEC T1TIME
T10   MOV T2TIME, T2TIME
       JEQ T20
       DEC T2TIME
T20   EQU $
       SBO CLOCK
       RTWP

*NOTE: T1TIME points to register 10 (R10) in
       each task workspace. Thus, the task
       itself also can look at this location
       easily to determine if the timer is
       at zero
         For Example: MOV R10,R10
                         JNE

as shown on *Figure 10-12.*

b. Coded Instructions

*Interrupt is caused by a hardware clock interrupt

a. Flow Diagram

When its timing cycle has been completed, the timer routine is automatically reset.

The timing continues decrementing counter values that are not zero and resetting interrupts. An interrupt service routine services the interrupt until the counter is zero, then the operation stops.

RAM memory locations for each operation control routine are loaded with the appropriate time value when the corresponding operation routine is initiated. The output of a constant frequency oscillator is connected to an interrupt line of the microcomputer. Each time the interrupt occurs, an interrupt service routine checks each counter value. If it is not zero, the value is decremented (1 is subtracted from the value) and the interrupt line is reset.

This sequence continues until the timer routine is reset by completion of the operation or the counter reaches zero. The operation control routine software also checks this software counter during its execution. If zero is read, a timeout error is indicated and the error handling software is executed to stop the operation.

LOAD STATION

Figure 10-14 shows a side view outline drawing of the load station. The operation is to load a flexible substrate from a supply maintained by the operator onto one of the boats for the assembly operation. Once a substrate is successfully loaded, the tolerances for the assembly operations are ensured by the original setup of the machine.

First, the vacuum head on the substrate pick device is raised to the bottom of the stack of substrates. A substrate is peeled from the bottom of the stack by the vacuum head, then the head and substrate are lowered. Next, the substrate shuttle mechanism moves the substrate pick device (to the right in *Figure 10-14*) until the substrate is under the pick-and-place device. This device is lowered and its two vacuum heads grab the substrate and lift it from the substrate pick vacuum head whose vacuum is now turned off.

Substrates are picked from a stack and loaded on a substrate all under control of software.

**Figure 10-14.
Substrate Load Station**

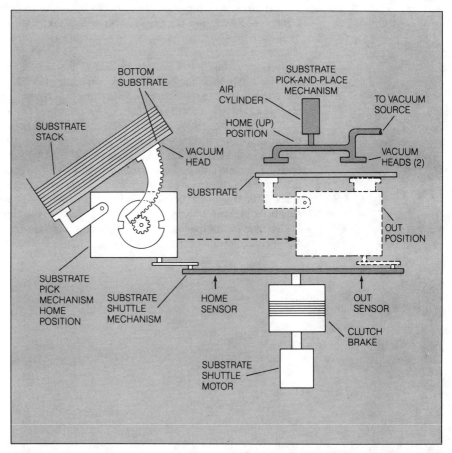

The substrate shuttle mechanism then moves the substrate pick device back (to the left) to its home position to obtain another substrate from the stack. As soon as it has moved back, the cross-slide shuttle of the transport mechanism (discussed earlier) moves an empty boat under the substrate pick-and-place device. This device lowers the substrate onto alignment pins of the empty boat to complete the load operation. The pick-and-place device then is raised to its home position to clear the boat and prepare for the next substrate.

This seemingly simple procedure requires the interaction of a number of system components. For example, the substrate pick device and the cross-slide shuttle mechanism must occupy the same space at different times during the process. Since the two mechanisms cannot be there at the same time, the control routine must time them properly to prevent a collision. All of the mechanical assemblies operate completely independently of one another, but they are synchronized by the software routine. To synchronize them with mechanical linkages would have required a very complex mechanical design using levers, gears and pulleys. The operation is quite complex and shows the power of software programmable control in reducing the complexity and number of mechanical controls.

Figure 10-15 shows the flow diagrams for the software that synchronizes the interaction between the substrate transfer and the main transport. Notice how the flags are set or cleared by one routine and checked by the other before advancing to the next step. The flag check routines run in a loop to constantly check the status of a flag. As soon as the flag is set or cleared as necessary, the control routine advances. Thus, independent software modules can be written for each operation which will run as fast as the machine will allow. In this case, the software is capable of running much faster than the mechanical hardware will allow; thus, this software is efficient.

The two stations which determine the cycle time limits of this machine are the substrate load station and the keypad attach station. Anything that can be done to reduce the cycle time of these two stations can increase the speed of the machine and make it more cost effective. One thing that will help is to have the substrate pick mechanism go ahead and pick a substrate from the stack while the boat is being loaded with the previous substrate. Then the substrate pick mechanism will be ready to move to the right as soon as the loaded boat clears the load station. By doing this, one-half second can be shaved off the transport time.

> The control system must be able to coordinate several activities at the same time, while preventing possible collisions or malfunctions between machine assemblies. While a complex and difficult problem to handle by mechanical control, a software routine can easily control all of these by a system of flag routines.

**Figure 10-15.
Substrate Transfer and
Main Shuttle Interaction**

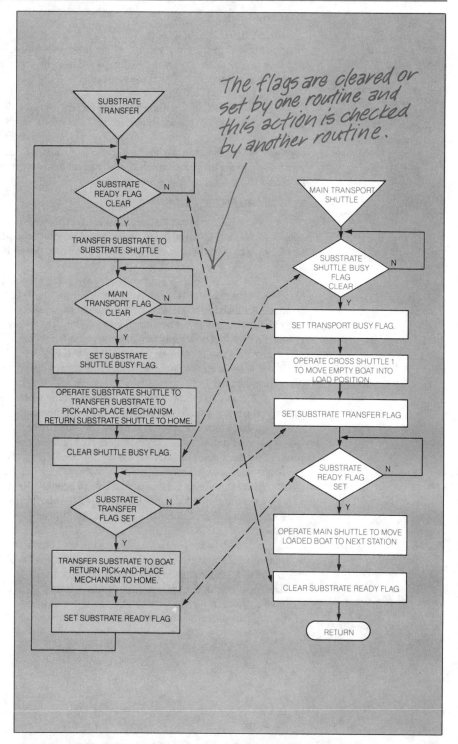

FLUX STATION

At this station, a solder flux which acts as a cleaning and wetting agent is applied to the substrate. The flux process requires an air cylinder actuator to extend and retract the flux pad. The cylinder is controlled by a solenoid valve and two position sensors indicate if the flux pad is home or extended.

The flow diagram and the coded instructions for the control program to run this station are shown in *Figure 10-16*. Notice that a flag is set by the routine just before returning to the main program to indicate that the operation has completed without an error. The routine is allowed to run by the control sequence program which keeps track of operator requests and monitors the machine for error conditions.

IC ATTACHMENT STATION

The IC attachment station is shown in *Figure 10-17a*. It requires two separate pieces of hardware; the IC material handler which allows the machine to carry a supply of ICs for extended operation and the pick-and-place mechanism.

The supply of ICs is carried by a revolving carousel unit which can handle several tubes of ICs so the operator doesn't have to reload the carousel so often. A mechanism known as an escapement is used for feeding. The lower end of this mechanism is shown in *Figure 10-17b*. The operation of the escapement is such that only one device can be fed from the carousel at a time and it provides the IC properly oriented for the pick-and-place mechanism. The carousel software ensures that an IC is always present for the pick-and-place mechanism.

A cam and follower powered by a stepping motor provide the action to pick an IC from the feed arm of the escapement mechanism and place the IC on the substrate. Once the IC is in place on the substrate, the heater element and a clamp are actuated by two air cylinders controlled by solenoid valves to provide the heat and pressure necessary for the solder reflow process. The pick-and-place arm, clamp and heater are shown in *Figure 10-17b*. The air pressure into the air cylinder is adjustable to provide the required mechanical pressure on the clamp for the solder process. Temperature and pressure sensors provide feedback for the control system to ensure proper station operation.

Two software control loops provide the control for the IC attach station; one for the carousel operation and one for the IC pick-and-place operation. The two loops operate independently and in parallel for faster machine operation. The interaction resulting from the parallel operation is handled by defining status flags so that each routine can determine the status of the other as described for the substrate load operation.

**Figure 10-16.
Flux Station Program**

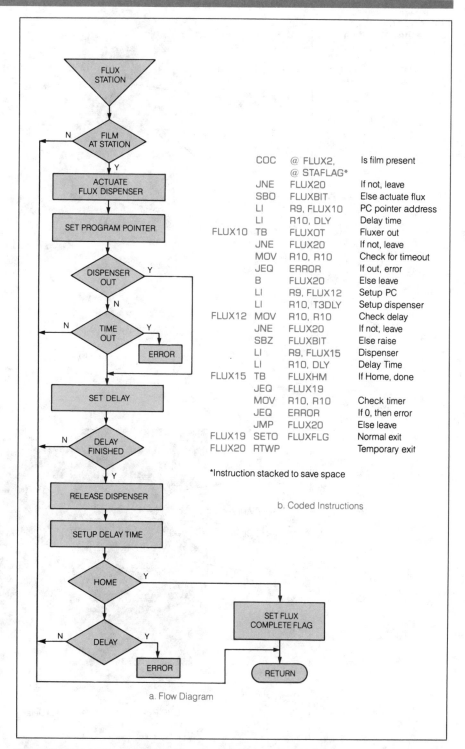

	COC	@ FLUX2, @ STAFLAG*	Is film present
	JNE	FLUX20	If not, leave
	SBO	FLUXBIT	Else actuate flux
	LI	R9, FLUX10	PC pointer address
	LI	R10, DLY	Delay time
FLUX10	TB	FLUXOT	Fluxer out
	JNE	FLUX20	If not, leave
	MOV	R10, R10	Check for timeout
	JEQ	ERROR	If out, error
	B	FLUX20	Else leave
	LI	R9, FLUX12	Setup PC
	LI	R10, T3DLY	Setup dispenser
FLUX12	MOV	R10, R10	Check delay
	JNE	FLUX20	If not, leave
	SBZ	FLUXBIT	Else raise
	LI	R9, FLUX15	Dispenser
	LI	R10, DLY	Delay Time
FLUX15	TB	FLUXHM	If Home, done
	JEQ	FLUX19	
	MOV	R10, R10	Check timer
	JEQ	ERROR	If 0, then error
	JMP	FLUX20	Else leave
FLUX19	SETO	FLUXFLG	Normal exit
FLUX20	RTWP		Temporary exit

*Instruction stacked to save space

b. Coded Instructions

a. Flow Diagram

**Figure 10-17.
IC Attachment Station**

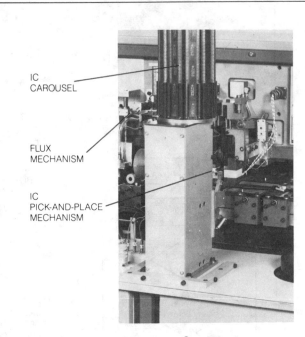

IC
CAROUSEL

FLUX
MECHANISM

IC
PICK-AND-PLACE
MECHANISM

a. Overall View

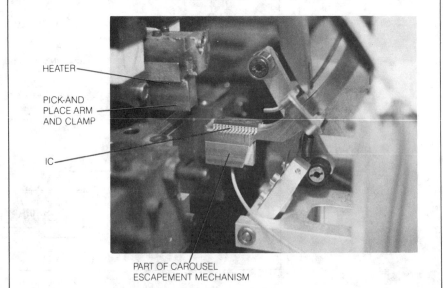

HEATER

PICK-AND
PLACE ARM
AND CLAMP

IC

PART OF CAROUSEL
ESCAPEMENT MECHANISM

b. IC Pick-And-Place Mechanism Detail

KEYPAD FLUX AND ATTACH STATIONS

These stations require the same steps as the IC attachment. The difference is in the size and shape of the keypad and how it is handled.

DIAGNOSTICS/TROUBLESHOOTING

Problems with process machinery can be found by running a diagnostic software routine that checks, either continuously or step by step, each software command for a particular operation.

Two things are done on this machine to help find machine faults. First, a separate software routine is provided which allows the operator to look at the status of all of the sensors on the machine by means of 16 status lights and a thumbwheel switch. Sensors are wired in groups so their outputs can be multiplexed to one input port of the controller. Each input port is identified with a unique address. By setting the thumbwheel to the correct port address as shown by the operator panel, the operator can review the status of up to 16 sensors available on that port. Changing the thumbwheel switch to another port address brings up the status of 16 different sensors. The software routine continually updates the indicated sensor status as the machine operates. The routine has been designed so it always runs, thus, it can be used while the machine is in normal operation or it can be used in a pure troubleshooting mode.

The second aid is the ability to select an operation mode in which only the operations at any one station of the machine are performed at a time. Besides normal operation of the selected station, a special step-by-step switch can be used to actuate the station one motion at a time. This provides the ability to stop the machine at critical points to check part alignment and to look for specific troubles. When in the diagnostic mode, some of the error routines are bypassed to keep them from stopping operation.

OPERATOR I/O

Errors found by the software diagnostic are displayed to the operator on a control panel, which also provides a means for inputting system parameters.

In addition to a control switch panel to operate the machine, a separate terminal is provided to display what error occurred; for example, if the machine goes through an automatic shutdown due to a time-out error. This terminal also provides the means to load the heating time, cooling time, etc. for the attachment process as well as provide production data on the number of assemblies built, run time, idle time, down time, etc. This information also is available at the central control point of the complete system and can be printed out so that a hardcopy can be maintained.

WHAT HAVE WE LEARNED?

1. Using a microcomputer as a controller in automation permits the mechanical portion of the design to be less complex and allows the actuator function to be changed in software rather than requiring redesign of the mechanics.
2. A process being considered for automation must be well understood and the system requirements well specified before the design of the automation system begins.
3. Other things to consider besides system requirements when choosing an automation controller are: cost, availability of maintenance service, programming language, and software support.
4. Flow diagrams of machine operation are very helpful in developing the program code for the control software.
5. A by-product of an automation system is the ability to monitor and report the status of the machine and various parameters regarding the assembled product.
6. When choosing the controller hardware, the most effective method is to select standard off-the-shelf components to integrate into a system. Advantages are: lower cost, shorter development time, proven reliability and compatible interfaces.
7. Generally, a microcomputer-based controller consists of a microprocessor or single-chip microcomputer, memory, I/O for machine control and peripherals, and a RS232 communications interface.
8. Structured software allows the program to be broken down into small modules with well-defined interfaces. This allows different programmers to develop software modules independently, yet all the modules will work together successfully. Often some of the modules can be used again in other applications with little or no change.
9. An operating system is a software program that keeps everything in the system operating together and in the correct order.
10. A function control block is a block of information that tells the operating system about a particular subroutine.
11. The operation control routine decides which application program to run at any given time. Application programs are those that control the machine to perform certain tasks.
12. Interleaved task operation improves the use of CPU time. Flags are used in interleaved operation to determine which portion of which task to run at any particular time.
13. A flag is a bit in a register or memory location that indicates the status of a particular condition by whether the bit is set to a 1 or reset to a 0.

14. A device service routine is an I/O interface between an application program and a peripheral such as a CRT or printer. It permits simplification of application programs since the application program only reads from or writes to specified buffers and the device service routine handles the rest of the I/O operation.

15. Machine errors are detected by software which checks the status of timers and sensor driven flags. If a particular flag is not set or cleared when it is supposed to be, or if an operation timer times out, an error is indicated.

16. In the example machine, the mechanical operations work independently, but were synchronized in software by status flag checking. This feature of software eliminates complex mechanical synchronization through levers, gears and pulleys.

17. Diagnostic software is very helpful in finding machine faults. The example machine had both a normal operation diagnostic mode as well as a step-by-step mode.

Quiz for Chapter 10

1. Most industrial control applications:
   a. can be solved using more than one method of control.
   b. require a careful problem description.
   c. can be done more reliably using solid-state methods.
   d. must have the process specified correctly to be successful.
   e. all of the above.

2. A way to transfer programs to a central system is:
   a. by setting bit switches on the central CPU.
   b. to copy the program into PROMs and install the PROMs in the system.
   c. to develop the program on the actual system only.
   d. any of the above.

3. An operating system:
   a. is required to run a microprocessor system.
   b. provides a structure with which to develop application programs.
   c. is useful because standard modules can be used.
   d. b and c above.
   e. all of the above.

4. An operating system:
   a. can help reduce software development time.
   b. is useless in most control applications.
   c. provides ability for other programmers to understand the software structure.
   d. all of the above.
   e. none of the above.

5. A favorable reason for automation is to:
   a. raise assembly cost.
   b. decrease throughput.
   c. perform difficult operations.
   d. increase number of operators.
   e. all of the above.

6. The program execution time:
   a. must be slower than the fastest mechanical operation that is to be controlled.
   b. will vary depending on the language used.
   c. cannot be changed once the program is written.
   d. all of the above.

7. Using a microcomputer as a controller allows:
   a. simpler mechanical hardware.
   b. faster cycle time.
   c. more self diagnostics of the machine operation.
   d. all of the above.

8. Standard computer control products:
   a. are available off the shelf.
   b. are very expensive and not cost effective.
   c. do not exist yet.
   d. none of the above.

9. Timing using a TM990 microcomputer:
   a. requires an external timer added by the user.
   b. is available on the standard CPU board.
   c. can be handled in software.
   d. b and c above.
   e. none of the above.

10. Computer-based controllers:
    a. should be built in a modular fashion wherever possible.
    b. are very difficult to change.
    c. are very flexible.
    d. a and c above.
    e. none of the above.

Future Trends and Reliability

ABOUT THIS CHAPTER

This chapter deals with the future direction of the use of microprocessor-based solid-state products for automation and with their reliability. Both topics are important when considering future commitment to automation.

REVIEW OF THE PAST

Studying the evolution of industrial automation control discloses that, at any particular point in time, a critical control problem was solved by using the most reliable, cost-effective method available at the time. The method chosen often did not utilize state-of-the-art technology because new technology typically is more expensive and has no proven reliability.

Devices using the latest technology are not always immediately incorporated into industrial designs when processes are being automated. The additional factors of cost and proven reliability have often led to more traditional approaches in equipment.

When computers were first introduced into automatic production equipment, the cost was high and the capability was limited compared to today. Memory especially was expensive and bulky. Magnetic core memory was the only random access memory available and its cost was about 50 cents per 8-bit byte. Presently the memory picture is much different because semiconductor integrated circuit memory is used extensively. It requires less space, and in production quantities, it costs less than 1/10 of 1 cent per byte. Because of the high cost and large space requirements of computers and their memory, the early machines were large central control processors that used assembly language programs.

DISTRIBUTED CONTROL

Advances in microcomputer technology make it more feasible to use distributed programmable controllers, instead of relying on central control facilities.

With the low cost and small size of integrated circuit memory and processors, and with the tremendous advances in the capability of the components, especially the programmable microcomputer, the central control processor system is no longer required. Now, the emphasis is on distributed computing, and programmable controllers are being introduced that employ distributed computing concepts. In the future, it will be an exception if distributed computing is not used.

Distributed control is where the controller and the control program are placed very close to the point being controlled. It is made possible by the increased functional capability of the integrated circuit components within a package. Microcomputers are available on a single chip in a single package that have as much capability today as a roomful of equipment just 20 years ago.

Interface components are available that not only handle digital inputs and outputs, but also include internal counters and timers, RAM and program ROM on the same chip. Some program ROM is large enough to contain programs in solid-state form that allow the system to run programs written in the high-level language BASIC.

As a result of bringing the controlling function closer to the controlled point, it is no longer necessary to run a bundle of sensor and actuator wires to a central processor; instead, only a few wires link the distributed control points to a central supervisory controller. In a pure distributed control system, the network links could be cut and the individual control loops would continue to function because all the control requirements for a particular function are local; that is, near the machine or process under control. The central supervisory controller is not involved in the actual control task, but schedules the tasks and provides system status and data to the process control management people and to the product manufacturing people. The only physical connection is a digital communications link from the central controller to the local controllers.

Figure 11-1 shows a typical industrial distributed control system. Ideally, the communications links, called buses, would use a protocol of signals called a standard bus structure such that any level of control could be added to any bus and all smart subsystems (explained later) could communicate with each other.

Another change that probably will happen soon is the use of fiber optic cables rather than metallic wires as the transmission link between the central controller and the distributed controllers. Fiber optic cables already are used in telephone systems and in other applications. They have proven advantages such as reduction of radio frequency and electromagnetic interference, high voltage isolation, and elimination of ground loops. The main requirement to make the use of fiber optics practical for industrial control is the development of fast low-cost optical receivers.

SENSORS AND ACTUATORS

Smart Sensors

A further extension of distributed control will be the use of smart sensors. As the capability of semiconductor components increases, the ability to combine analog and digital processing on the same semiconductor substrate will permit the sensor output to be in digital form. This will increase the sensitivity and noise immunity of the sensor and its coupling to the controller. This is already being done on many digital voltmeters, for instance, where a complete meter—from analog voltage input to display control—is accomplished on one integrated circuit. This capability will be expanded further in industrial control to allow programming of the chip for a specific application and sensor input.

Distributed control has only a few control lines running from the controlled points to the central processor. In addition, the individual distributed control units continue to operate if the central monitoring computer goes out.

Smart sensors will have A/D and D/A processing built into the sensor. Further development will include programmability.

**Figure 11-1.
Distributed Control
System**

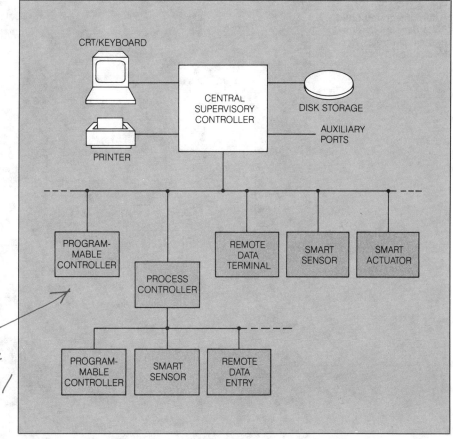

*The componets
required to do
the actual
controlling
are located at
the individual
process site,*

Solid-State Sensors

Dew Point Sensor

Adding a microprocessor for calculations to other solid-state devices should result in a dew point sensor that can be read directly.

Solid-state sensors such as the strain gauge pressure sensor mentioned in Chapter 3 have begun to appear over the last few years and the available types will continue to increase. One now being evaluated is a solid-state psychrometer which is used to measure the dew point of an atmosphere. The sensor consists of a solid-state cooling device, a temperature sensing element and a capacitive moisture sensing element. As the device cools, moisture condenses at the dew point and the capacitance changes. This signals the system to sample the temperature sensor output to calculate the temperature. In the future, a microprocessor could be added to the sensor package to calculate the actual dew point temperature. The digital output of this smart sensor would be linked to the controller by a fiber optic cable. The new method is contrasted with the old method in *Figure 11-2*. The sensor is called a "smart" sensor because it has the ability to calculate the dew point temperature and provide this data in a standard format that can be read directly by an operator or another computing element.

**Figure 11-2.
Future Integrated Solid-
State Sensors Versus
Present Discrete
Mechanical Type**

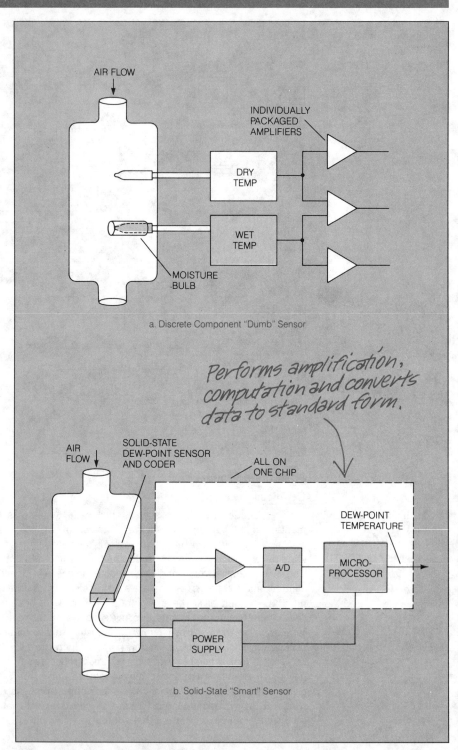

AIR FLOW

INDIVIDUALLY
PACKAGED
AMPLIFIERS

DRY
TEMP

WET
TEMP

MOISTURE
BULB

a. Discrete Component "Dumb" Sensor

*Performs amplification,
computation and converts
data to standard form.*

AIR
FLOW

SOLID-STATE
DEW-POINT SENSOR
AND CODER

ALL ON
ONE CHIP

DEW-POINT
TEMPERATURE

A/D

MICRO-
PROCESSOR

POWER
SUPPLY

b. Solid-State "Smart" Sensor

Gas Flow Sensor

Measuring the flow rate of a gas is another application for a solid-state sensor. As shown in *Figure 11-3*, three transistors on a common substrate are placed in the gas flow. The two outside units are temperature sensors and the center unit is the heater element. The difference in the temperatures measured by the two sensors is a function of the gas flow. An integral microprocessor can use a built-in program to calculate the gas flow so the output of the solid-state sensor can be a digital bit stream giving the gas flow in the absolute units used by the control engineer.

**Figure 11-3.
Solid-State Gas Flow
Sensor**

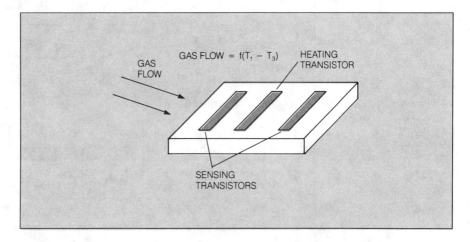

Smart Actuators

A smart actuator with computational capability will do all the controlling after receiving a set point from the central monitoring computer.

Analog control actuators presently use a current loop to carry the proportional control signal from the controller to the actuator. Additional analog control is used at the actuator to transform the current loop signal to actual movement of the actuator.

By using digital techniques with a programmable microprocessor integral with the actuator, the position response of the actuator to an input signal can be calculated and the data transmission can be changed to a serial digital bit stream. The programmable controller need give only position information and the smart actuator will solve the remainder of its control loop and position itself.

Furthermore, if a smart actuator could sense its own position, it could respond directly to an absolute command from the central processor. For instance, in a conventional system, if the controller wanted to obtain a certain flow rate, it would drive the actuator flow control while monitoring the flow rate sensor until the desired flow rate was obtained. By interfacing a smart actuator directly to a smart sensor, the controller need only give a flow amount to the actuator. The calculations and flow control adjustment would be handled directly by the smart actuator. Thus, the central controller's only function would be to transmit desired conditions and monitor the status of the peripheral unit. *Figure 11-4* shows the difference between present actuator interfaces and those which will be used in the future.

**Figure 11-4.
Actuator Interface**

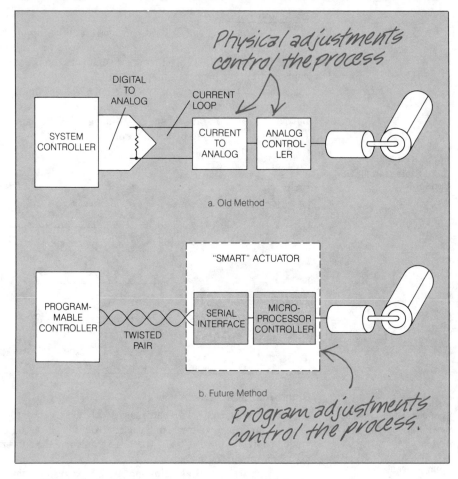

Physical adjustments control the process

DIGITAL TO ANALOG

CURRENT LOOP

SYSTEM CONTROLLER

CURRENT TO ANALOG

ANALOG CONTROL-LER

a. Old Method

"SMART" ACTUATOR

PROGRAM-MABLE CONTROLLER

TWISTED PAIR

SERIAL INTERFACE

MICRO-PROCESSOR CONTROLLER

b. Future Method

Program adjustments control the process.

Data Transmission

Smart actuators and sensors could be addressed and read from or written into just like memory locations. Data transmission would be simplified and error correction added.

Smart sensors and actuators could be given an address and be addressed just like a memory location. This would allow all of them to share one common data bus for transmission of data and commands. The controller would transmit the address of the specific actuator desired along with the data for that actuator. That actuator, but no other, would "turn on" its data receiver to accept the data. A sensor would be addressed similarly whenever the controller wanted its output. This addressing technique would greatly reduce the amount of wiring in the control system.

Since the outputs of these sensors and inputs to these actuators will be a digital bit stream, it will be possible to perform operations on the bit stream to detect errors produced in transmission. Not only will it be possible to detect an error, but also the error could be corrected by using a cyclic redundancy code (CRC). The CRC is a set of additional bits uniquely related to a block of data that is ready to be transmitted and the CRC is then transmitted with the block of data.

At the receiver, the same method is used to generate a CRC based on the received data. This CRC is compared with the CRC transmitted; if they do not match, an error is indicated. By reconstructing the data with a simple algorithm, it is possible to recover the data in its correct form even if several errors occur. This technique is already used in some data communications applications and its application in industrial control will ensure more reliable transmission even under electrically noisy conditions.

PROGRAMMABLE CONTROLLER

The capability of future programmable controllers will increase greatly. They will be made easier to use and more flexible. The addition of smart peripheral modules which can be programmed and controlled with the main controller program will greatly increase the design options of the control engineer.

The ability to perform several tasks or the ability to control a number of processes at the same time will make the programmable controller look more like a microcomputer. The difference will be in the dedicated nature of the language used with the controller. The programmable controller will be much easier to use since each controller will be dedicated to the solution of a specific industrial control problem. The ability to create even smarter controllers by adding smart peripherals will make this trend toward dedicated controllers grow in popularity. *Figure 11-5* summarizes the evolution of industrial control.

Although programmable controllers will function more and more like microcomputers as smart sensors, actuators and peripherals evolve, they will still be dedicated systems with their own unique language.

**Figure 11-5.
Evolution of Industrial
Control**

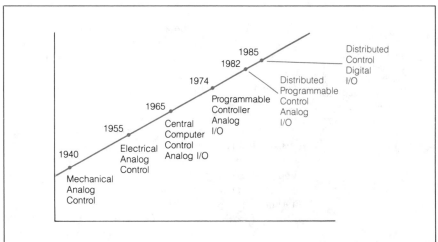

The movement is toward programmable processing closer to the point of control.

Trend Analysis

Trend analysis is the analysis of the history of a variable in a process. When trend analysis is included in the control loop decision, a much better decision can be made because the data includes past history instead of only current data. Trend analysis has been available in control systems using large computers, but not in systems using low-cost controllers. With the increased computing power of the microcomputer-based controllers, this important analysis will be available in many more systems.

Predicting preventive maintenance is an application that can use the results of trend analysis. By observing the trend of a characteristic parameter of the component, it is possible to forecast when a process component will go out of specification. The maintenance person can use this information to schedule the time to make adjustments or replace parts to prevent the out-of-specification condition from ever occurring.

MICROCOMPUTER CONTROL SYSTEMS

Standardization of software language for industrial application will provide a significant boost to the use of microcomputers in control systems.

Microcomputers for industrial control continue to include more functions in a given space with more peripheral I/O boards and more system configurations. The main thrust in the future will be to make computers easier to use for industrial control. At present, the main limitation to their use is the development of the software to accomplish the needs of the particular application. As more systems are put into use, the need for a standard industrial control programming language and an operating system becomes more apparent. This will provide a common software interface which can run programs written by many users.

Once a common language is developed, the next major step will be the use of speech recognition equipment so that the programmer and operator can literally talk to the computer. The combination of easy to use operating systems and easy to input control programs will greatly expand the use of the microcomputer in the industrial control environment.

ROBOTS

Sophistication in robot capabilities will only increase their greatest assets, flexibility and variety of use. Increases like this will lead to a more extensive employment of robots and hence, a lower cost.

Today, most robots are used for pick-and-place applications and this form of automation will increase dramatically in the next decade. Many applications that now use standard automated equipment will be converted to use robots. As mentioned in Chapter 9, the more sophisticated robots have the capability to learn an operation by the "walk-through" method with a skilled human teacher so that very little actual programming is involved. This makes their use much easier than the standard automated machine.

One main advantage of robots is that the same robot system can fulfill entirely different purposes. This is different from the automated machine which is dedicated to a specific purpose and is the primary reason that the robot is so exciting. It is not that it does something that no other machine can do, but rather that the same machine is not limited to just one task in its lifetime. The key is flexibility.

Flexibility allows the development and overhead costs of the robot system to be spread over a much larger number of machines, thus lowering the individual robot cost per unit of manufactured product. This in turn feeds on itself for it makes the robot affordable in more applications, which increases the number of robots required. The increased number of robots built permits the robot manufacturer to lower the cost of each robot leading to more applications, etc.

Robot Eyes

One of the main requirements of a robot which will greatly increase its effectiveness is the use of image sensors to give it the ability to "see". A robot without vision must be guided by very precise dimensional and angular inputs to perform a pick-and-place task. A robot with vision can be given directions to position its arm to the approximate location, then it can use its image sensors to guide itself to the precise location.

A TV-type camera or solid-state imager is used as a vision sensor in some robot applications to give two-dimensional vision, but a large amount of computer time and a large memory are required to process the image. Of course, bilateral sensors required for three-dimensional vision compound the computation and memory problem. Even the most sophisticated robot vision available today cannot come close to the capability of human vision. However, here is where much development will be concentrated to obtain image sensors which can more closely mimic the vision capabilities of the human. This vision development coupled to greater computational power and memory in smaller and less expensive packages will provide robots capable of fantastic feats.

RELIABILITY OF AUTOMATION EQUIPMENT

When using new technology components there is a reliability risk involved because the product does not have a reliability history.

In order for a product to be accepted readily, it should have a history of reliable operation. A newly developed product, even though it incorporates a better technology and ultimately may prove to be more reliable, initially has no history. Therefore, to use a component or product for the first time involves a certain amount of risk. In many cases, it may be better to stay with a component or product that has a proven record than to take the risk. Even the latest advance into space, the space shuttle, contains equipment developed in the 1960's because it has a proven reliability.

The evaluation of the risk when replacing mechanical or electromechanical parts with solid-state components involves the following:

1. The wear on mechanical or electromechanical parts usually causes reliability problems. Either preventive maintenance or repair downtime must be considered for the machine.
2. Electronic control does not eliminate completely the mechanical or electromechanical parts, but reduces the number significantly and improves reliability accordingly. As a result, downtime is reduced and time between preventive maintenance is increased greatly.
3. When proper design margins are included, most electronic failures or machine faults can be traced to solder connections or cable connections. Using integrated circuits where all components are made at the same time on a slice of silicon with uniform and standard processing, the interconnection by solder connections and cables is greatly reduced. Orders of magnitude improvement in reliability occur because of this fact.

4. Even though a new electronic control has not been used previously, the reliability of the individual components (which have an excellent history because they have been manufactured for some time) provides an excellent foundation for a predicted reliability of the new control.
5. The reliability of a new family of integrated circuits may be predicted based upon the proven reliability of similar types of integrated circuits made with the same process. Thus, the reliability of the new product which uses the new IC also can be predicted to provide confidence in using the new product in a system.

Failure Rate

By plotting the life cycle of a component, a component's most stable/reliable period can be determined, as well as when it is likely to fail. The period of least stability and greatest failure rate is at first, the infant mortality period.

Figure 11-6 shows the typical failure rate versus operating time of a component. (This is often referred to as the bathtub curve for an obvious reason.) The curve shows a period of relatively high failure rate early in the component's life, then a period of constant rate of failure, and finally a gradual increase in failure rate near the expected end of its life. The early high failure rate period is referred to as the infant mortality period. The period of constant failure rate is the portion of a component's life that is most useful. Finally, a preventive maintenance plan would require component replacement before the time of increased failure rate to prevent equipment failure due to the component failure.

Semiconductor devices have a very short infant mortality period of only about 50 to 200 hours of operation. After this period, they have a very long period of constant, but low, failure rate. Thus, if each semiconductor device were operated in a test circuit for the length of the infant mortality period and it did not fail, it could then be installed in a manufactured product with confidence that it would not fail until it reached its expected end of life. This procedure would greatly increase the reliability of the manufactured product.

Figure 11-6.
Failure Rate Versus
Component Operating
Time

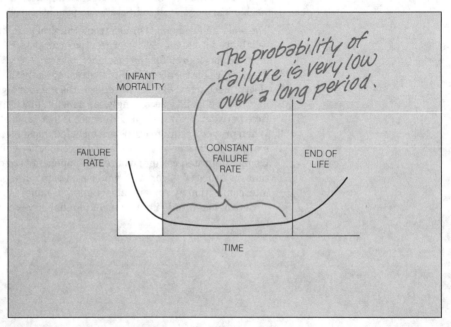

In fact, this method of eliminating infant mortality failures, called burn-in, is used for many solid-state products.

In spite of this, the use of solid-state components to replace electromechanical ones in industrial control applications at first met resistance from control engineers because they did not see any significant benefit in changing their controllers. However, they are changing their views as the much greater reliability of solid-state controllers is proven.

Mean-Time-Between-Failures

The mean-time-between-failures is the average time that a component, circuit, subsystem or system will operate satisfactorily before failing. It provides a reasonable means of comparison of actual operating reliability, however, all calculations are based on tests at worst-case conditions.

One measure of reliability is the Mean-Time-Between-Failures (MTBF) which is the average time, usually expressed in hours, that a component or system will operate from one failure to the next. The MTBF is a statistical average obtained by testing a meaningful sample of parts and, with some specified level of confidence, predicting the MTBF based on the test results. If the failures that occur are random in nature, the curve showing number of failures versus time would be that of a normal distribution (the familiar bell-shaped curve); that is, a few parts fail early and a few fail late, but most fail within a narrow range about the same time.

The conditions under which MTBF is measured should be the worst case. This means the component should be subjected to both high and low temperature extremes, high relative humidity, vibration, shock, and possibly a salt spray atmosphere. Of course, these conditions must not exceed the manufacturer's ratings. The MTBF of a product when not stressed to the limits of its operating capability normally will increase tremendously. Thus, if the product actually will be operated in an air-conditioned environment where humidity and temperature are controlled and there is little shock and vibration, it will have a much higher MTBF than if it is used under worst case conditions. Of course, MTBF also can be increased by an effective preventive maintenance plan.

Since a new product has no operating history, its anticipated reliability must be determined by other means. This can be done using modeling techniques and by projecting it from the reliability of similar product types. MIL-STD Handbook 217 also gives recommended methods for computing a reliability figure for semiconductor components under worst case stress conditions. Usually these will yield a reasonable prediction of MTBF. They also can be used to compare reliability statistics when, for instance, the product is used at other than the specified temperature.

Mean-Time-To-Repair

The time required to diagnose and repair a failure (the mean-time-to-repair, MTTR) is another factor used for reliability comparisons. It can affect the inventory level of spare parts significantly.

The mean-time-to-repair (MTTR) is the average time to diagnose and repair a failed system. This will depend on a number of factors; the ability to easily troubleshoot the system, the availability of replacement parts, the experience of the maintenance personnel with the particular product, and the method of how the component will be repaired. If an entire system can simply be unplugged and a spare system put in its place, then little troubleshooting or repair time is required. However, a spare system must be maintained as a backup. At the other extreme, if the repair is to be done at the actual failed part level, troubleshooting time will be much longer. The tradeoff is in the cost of the spare parts that must be maintained in inventory versus the cost of loss production due to down time. If the proper level of repair is chosen by calculation or through analysis of actual failure data, it should be possible to reach a compromise so that only a relatively small inventory of the parts that cause most of the failures must be maintained.

Product Availability

Product availability (PA) is a more meaningful statistic, especially when the system is constructed primarily of semiconductor parts. It is a ratio between the total life time of a product and the time it is actually available for use. It is defined by:

$$PA = MTBF/(MTBF + MTTR)$$

When using semiconductor based components, the products availability (PA) is a more accurate judge of the cost trade-offs involved between equipment failure, parts inventory, repair methods, etc.

In the past, when equipment was of a simple nature, MTBF alone could be used to judge the availability of the equipment. Failures were fairly frequent, but the repair time was short. The control system was simple and easily understood and the maintenance people could quickly diagnose many problems because the same problems occurred over and over at short intervals.

The use of semiconductor components has changed this because the MTBF typically is a large number. When a failure does occur, the maintenance personnel usually don't have as much experience in troubleshooting that particular equipment; therefore, the MTTR can go up quite a bit. The following example illustrates the difference.

A control system made up of twenty relays of the same type is in operation for one 8-hour shift a day. The MTBF for one relay is 1,000 hours of operation; however, the MTBF of the system is calculated to be approximately 50 hours; that is, a system failure is expected about every six days of operation. The maintenance person will become quite proficient at troubleshooting and replacing the relays after a short period of time. If the MTTR is assumed to be two hours, the product availability is 96% of the time.

If the same system uses solid-state components, let's assume the system MTBF is increased by a factor of 10; thus, the time between failures is more than 60 days of operation. The machine is more reliable as expected; however, it takes more time to troubleshoot. In fact, to troubleshoot down to the component level now takes 20 hours which is also ten times as long as before. The product availability has not changed, it still is 96% (480/480 + 20). However, to the equipment user it may appear to be worse because it is down for 20 hours at a time. This could be a major sore point with a production planning organization because it may be easier to allow for a recurring two-hour per week down time than a somewhat unpredictable 20-hour down time that might occur just when maximum production is needed.

Maintenance Method For Solid-State Controllers

Field maintenance programs and troubleshooting techniques for solid-state controllers should emphasize internal fault detection routines and troubleshooting and repair at the sub-assembly, rather than component level.

The major point of the above discussion is that a different maintenance program is required for solid-state controllers. Built-in fault indicators, self-test routines and step-by-step operation are desirable to reduce troubleshooting time. Preventive maintenance routines that try to pinpoint a parameter that is approaching a limit also are very important.

Troubleshooting techniques also change. Instead of repairing down to the component level when a fault occurs, it will be better to maintain an inventory of spare high-level subassemblies. When a fault occurs, a full subassembly is replaced. If the equipment is not manufactured in-house, an exchange program with the manufacturer or equipment supplier would provide a new or reconditioned subassembly for a service fee. Troubleshooting the defective subassembly to a component level is handled by the equipment manufacturer who uses the same type of extensive automated test equipment and experienced personnel that was used during manufacturing tests of the product. If this approach is used, it is even more important than usual to select a reputable, reliable manufacturer who will be in business for the expected life of the product.

CONTROLLER RELIABILITY SPECIFICATIONS

The environment and conditions under which the controller is to operate are extremely important factors to be considered when evaluating a device's reliability and potential use.

Each application, of course, must be evaluated on an individual basis. Certainly the particular environment in which the product is used may be very different from that which is specified on the manufacturer's data sheet. However, for many applications the environment is not as severe as the worst case conditions in which the equipment is tested by the manufacturer. For these applications, the equipment usually will be more reliable.

When choosing a product for use in a control system, first determine the major factors that affect reliability. Major factors in many cases are environmental factors such as the operating temperature range, humidity, shock and vibration. If operating or environmental conditions are extreme, then the product chosen should be a ruggedized product that has been designed and tested to operate at the extended limits or in the extended environment. Product life of the system can be extended if a product is chosen that will operate at limits that are less severe than those used for reliability tests. This also helps to reduce failures. If a product is going to be operating close to its reliability limits, try to choose a product which will allow preventive maintenance procedures to be performed easily to reduce system down time.

Microcomputer Controllers

Microcomputer systems used as industrial controllers in many cases were not specifically designed for a harsh environment, yet they have a good reliability record. The TM990 modules used in the calculator assembly example discussed in Chapter 10 have a demonstrated MTBF of over four years. These products are available with a conformal coating over the surface which seals the components and printed circuit board to protect them from moisture in a high humidity environment. The coating also provides some mechanical support to components which is of value if the equipment is subjected to vibration. As the TM990 products are used more and a history is built, design improvements can be incorporated to further increase the MTBF.

Figure 11-7 contains some data on TM990 board products subjected to reliability testing at 65°C. Only nine component failures occurred during a total of 376,000 board operating hours. The MTBF was calculated to be over four years. *Figure 11-7* indicates that a dynamic burn-in which is done on all of the board products before the start of the reliability test significantly reduces infant mortality.

The excellent reliability records of solid-state systems is shown in the TM990, a microcomputer board product which has a MTBF of over 4 years. PC controllers are designed for even harsher environments and to use special troubleshooting techniques.

**Figure 11-7.
TM990 Reliability
Test Data**

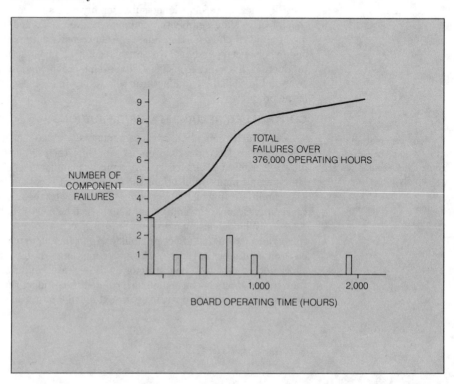

Programmable Controllers

Programmable controllers are designed specifically to operate in a harsh industrial environment and burn-in usually is performed to reduce field failures. MTTR is reduced by purchasing spare subassemblies to quickly repair the system. Special system troubleshooting analysis quickly identifies the defective module or replaceable subassembly. The subassembly is then repaired in an off-line facility or sent back to the manufacturer for repair. With such procedures, MTTR is short and MTBF is long, thus product availability is very high.

WHAT HAVE WE LEARNED?

1. In the future, distributed computing will be used more throughout the automation industry.
2. More data communications networks with standard bus architecture will be used to provide better control and reporting of the manufacturing and process operations.
3. As more electronic control is applied in automation systems, equipment cost will decrease which will make it cost-effective for more manufacturing processes to be automated.
4. The performance of sensors will improve due to the use of solid-state components to perform the actual sensing operation. Many will include a microprocessor right on the same chip.
5. Actuators will have more intelligence at the point of control and will require less interface to and computation by the control processor.
6. While many robots are relatively unsophisticated and others are much too expensive for most applications today, the robot has great potential for the future. They definitely will become more affordable and more useful.
7. Developing vision sensors and vision processors which are cost-effective probably will be the greatest challenge in expanding the capabilities of the robot.
8. More functional capability will be added to programmable controllers while maintaining the ease of programming tied to the ladder diagram. They will continue to provide a reliable, low-cost easy-to-use solution to many automation applications.
9. The use of microprocessors and microcomputers for industrial control will become more widespread as a standard automation operating system and easy to use languages are developed.
10. The use of semiconductor-based equipment will increase as more people recognize its advantages and better reliability.
11. New maintenance techniques must be used to reduce MTTR as solid-state control equipment with much longer MTBF tends to have longer MTTR.

Quiz for Chapter 11

1. Historically, central systems were used because:
   a. memory was very expensive.
   b. semiconductor memory was not available.
   c. memory consumed a large amount of power.
   d. all of the above.
   e. none of the above.

2. Sensors:
   a. probably will not improve over the next decade.
   b. are easily developed.
   c. are not important in automation.
   d. will be much smarter and more sophisticated in the future.
   e. all of the above.

3. A smart solid-state sensor is:
   a. one that integrates both logic and a solid-state sensor into one unit.
   b. one that has two or three substrates.
   c. not available today.
   d. all of the above.
   e. none of the above.

4. Most data transmission in the future will be:
   a. digital bit stream encoded data.
   b. analog in nature.
   c. carried over metallic telephone lines.
   d. none of the above.

5. VLSI logic:
   a. means more costly electronic products in the future.
   b. will not be available in this decade.
   c. will result in lower cost, higher capability electronic equipment.
   d. all of the above.

6. Trend analysis:
   a. is used only in financial planning.
   b. can improve control system response.
   c. is a brand new concept.
   d. requires a very expensive system to implement.
   e. none of the above.

7. Microcomputer-based control systems:
   a. will just keep their present capability.
   b. will have more standard industrial application software available in the future.
   c. will tend to increase in cost.
   d. all of the above.
   e. none of the above.

8. Speech recognition:
   a. will never be used in industrial controls.
   b. is already widely used in industrial controls.
   c. will be used in more applications as the technology matures.
   d. none of the above.

9. The robot:
   a. represents the most powerful new industrial automation tool currently on the horizon.
   b. is very primitive at the present.
   c. will be more useful when robot vision is improved significantly.
   d. all of the above.

10. Vision sensors:
    a. can't improve much in the next decade.
    b. will remain very costly.
    c. will more closely match the capabilities of the human eye.
    d. all of the above.

11. The reliability life cycle of a product:
    a. has three distinct rates of failure.
    b. cannot be predicted.
    c. stays constant with environmental conditions.
    d. all of the above.
    e. none of the above.

12. Semiconductor devices:
    a. are less reliable than electromechanical components.
    b. usually have a very short MTBF.
    c. have a very short infant mortality period.
    d. all of the above.
    e. none of the above.

Glossary

Actuator: A device which performs an action in response to an electrical signal.

Amplifier: A device used to increase signal power or amplitude.

Analog Control: A type of control whose control variable varies smoothly and continuously in amplitude.

Analog-to-Digital (A/D or ADC): A device that converts a sampled analog signal to a digital code that represents the sampled analog signal.

APT: Automatically Programmed Tools language that is used to control the positioning, movement and contouring of a machine tool.

Assembly Language: An abbreviated computer language using mnemonics which humans can use to program computers.

Automation: The use of electrical, electronic and/or mechanical power to control automated processes.

Batch Process: A process where some operation is performed on one or more inputs in order to produce a finished product.

Binary Number System or Code: A method of writing numbers by using two numeral digits, 0 and 1. Each successive bit position in a binary number represents 1, 2, 4, 8 and so forth.

Bit: A binary digit; the smallest piece of data a computer can manipulate.

Byte: A group of 8 bits that are manipulated as a group.

Central Processor Unit (CPU): Part of a computer system which contains the main storage, arithmetic unit, and special register groups. It performs arithmetic operations, controls instruction processing, and provides timing signals.

Closed-Loop System: A control system in which the output is fed back and compared to the input to generate an error signal. This error signal is used to generate the new output signal.

Computer Numerically Controlled (CNC) Machine: A machine tool that is controlled by a computer dedicated to that machine.

Continuous Control: The ability to continuously change an input parameter from full off to full on.

Continuous Process: A process where raw materials enter one end of the system and the finished product exits the other end.

Controlled Variable: The process variable regulated by the process control loop.

Controller: The element in a process control loop that evaluates the controlled variable error and initiates corrective action by a signal to the controlling (manipulated) variable.

Dead Time: The delay time caused by physical distance before any change in the manipulated variable is felt by the sensor.

Digital Circuits: Electronic circuits whose outputs can change only at specific instances and between a limited number of different voltages.

Digital Control: This use of digital circuits and digital techniques to solve the control application.

Digital Signal: Information in discrete or quantized form; not continuous. Binary digital signals have one of two states (0 or 1) defined by voltage or current levels.

Digital-to-Analog (D/A or DAC): A device that converts a digital code into its equivalent analog signal level.

Direct Numerically Controlled (DNC) Machine: A system whereby a computer is used to control two or more machine tools.

Disturbances: Disturbances are parameters which affect the controlled variable, but which are not capable of being controlled by the controlling processor.

EAROM: An Electrically Alterable Read-Only Memory. Sections of memory with this device can be electronically changed.

EPROM: An Erasable Programmable Read-Only Memory that can be erased and reprogrammed.

Execution: The phase of a computer instruction cycle during which the instruction operation is actually performed.

Fetch: The phase of a computer instruction cycle during which the next instruction to be executed is obtained from system memory.

Flip-Flop: An electronic digital device that stores one binary digit (bit) of information as a 0 or a 1.

FORTRAN: A complex high-level computer programming language that is mathematically based and oriented for the solution of scientific problems.

Frequency Response: A graph of a system's response to different frequency input signals. (See Transfer Function.)

Gain: The ratio of a system's output magnitude to its input magnitude.

Hexadecimal Numbers: Numbers whose base is 16. The letters A-F represent the decimal numbers 10-15. The binary number 1001 1101 1111 0111 is 9DF7 in hexadecimal.

Integrator: An electronic circuit that performs the mathematical operation of integration. The output signal is proportional to the area under the input signal waveform when the input signal is plotted against time.

Interrupts: An efficient method of quickly requesting a computer's attention to a particular external event.

Ladder Logic Diagram: The diagram that is used to describe the logical interconnection of the electrical wiring of certain control systems.

Lead Term: A control system component which anticipates future inputs based on the present trend of a signal.

Limit Cycle: A mode of control system operation in which the controlled variable cycles between extreme limits with the average near the desired value.

Load Variable: A change similar to a disturbance but which is expected by the nature of the controlled process.

Logic Circuits: Digital electronic circuits commonly called gates which perform logical operations such as NOT, AND, OR and combinations of these.

Machine Language: The lowest level programming language. The coded instructions consist of a string of binary digits.

Manipulated Variable: The manipulated variable or variables are those parameters which are changed by the controller in order to maintain the controlled variable at the desired value.

Memory: A computer subsystem used to store instructions and data in the form of binary codes.

Microcomputer: A physically small computer which uses a microprocessor as a CPU and contains all the functions of a computer.

Microprocessor: An integrated circuit containing all the CPU functions.

MTBF: Mean Time Between Failures. The average time, usually expressed in hours, that a component or system will operate from one failure to the next.

MTTR: Mean Time To Repair. The average time to diagnose and repair a failed system.

Numerically Controlled (NC) Machine Tool: A power driven tool or set of tools whose actions are controlled by a computer program contained on a paper or magnetic tape.

Open-Loop System: A control system whose output is a function of only the inputs to the system.

Operational Amplifier: A standard analog building block with two inputs, one output and capable of very high voltage gain.

Optimal Damping: The damping which produces the very best time response.

Parallel Data: The transmission or processing of an n-bit binary code n-bits at a time.

Peripheral: An external input-output device which is connected to a computer.

PI: The designation of a controller operating in the proportional-integral mode combination.

PID: The designation of a controller operating in the proportional-integral-derivative mode combination. Also called three-mode.

Process: Any system comprised of dynamic variables, usually involved in manufacturing and production operations.

Process Variable: Any process parameter which can change values.

Processor: See CPU.

Program: A set of instructions used by a computer to accomplish some task.

Program Language: A set of rules governing the form used to write instructions.

Programmable Controller (PC): A programmable solid-state device capable of controlling a process or machine.

PROM: A Programmable Read-Only Memory device.

Proportional Amplifier: A control system component which produces a control output proportional to its input.

Proportional Control Mode: A controller mode in which the controller output is directly proportional to the controlled variable error.

Rate Action: Another name for the derivative control mode.

Read-Only Memory (ROM): A type of memory whose locations can be accessed directly and read, but cannot be written into.

Relay Ladder Logic: See Ladder Logic Diagram.

Reset Action: Another name for the integral control mode.

Robot: A programmable manipulator designed to move material, parts or tools through variable motions.

RTD: Resistance temperature detector. A copper, platinum or silver element whose resistance varies linearly with temperature.

Sampling: The act of periodically collecting or providing information about a particular process.

Semiconductor: A material which is neither a good conductor nor a good insulator. The basic material for all transistors, diodes and integrated circuits.

Semi-continuous System: A system where a sequence of operations is performed and where each operation of the sequence is continuous.

Sensor: An energy conversion device which measures some physical quantity and converts it to an electrical quantity.

Serial Data: The transmission or processing of an n-bit binary code in sequence one bit at a time.

Setpoint: The desired value of a controlled variable in a process-control loop.

Signal Conditioning: Changing or altering the output of I/O devices so that signals are in a form to be acted on by electronic circuitry.

Software: The computer program instructions that tell a computer what to do.

Steady-State Error: The difference between the output of a system and the input into the system after the input has been applied for a relatively long period of time.

Strain Gage: A transducer that converts information about the deformation of solid objects, called the strain, into a change of resistance.

System Lag: The total amount of delay time in a system from the time a manipulated variable is changed until the controlled variable responds.

Thermistor: A temperature transducer constructed from semiconductor material that converts temperature change into resistance change.

Thermocouple: A junction between two dissimilar materials that produces an almost linear voltage as a function of the temperature of the junction.

Time Constant: A number characterizing the time required for the output of a device component, subsystem or circuit to reach approximately 63% of the final value following a step change of its input.

Transfer Function: The response of an element of a process-control loop that specifies how the output of the device is determined by the input.

Transfer Lag: The time it takes for the manipulated variable to have an effect on the process.

Transient Response: The immediate response of a system to a change of input.

Venturi: A type of restriction used in flow systems to facilitate measurement of flow by the pressure drop across the restriction.

Index

Answers to Quizzes

Chapter 1
1. b
2. d
3. b
4. d
5. d
6. b
7. d
8. b
9. c
10. c
11. b
12. a
13. d
14. b
15. b
16. d
17. a
18. d
19. d
20. a

Chapter 2
1. c
2. d
3. d
4. d
5. b
6. d
7. d
8. d
9. b
10. d
11. c
12. d
13. e
14. a
15. d
16. b
17. d
18. d
19. d
20. b

Chapter 3
1. e
2. d
3. c
4. d
5. d
6. c
7. e
8. d
9. d
10. d
11. e
12. b
13. d
14. d
15. d
16. d
17. e
18. d
19. c
20. d

Chapter 4
1. d
2. b
3. d
4. a
5. d
6. a
7. c
8. d
9. c
10. b

Chapter 5
1. b
2. d
3. a
4. b
5. b
6. a
7. a. 1
   b. 3
   c. 4
   d. 2
8. a
9. d
10. c
11. b
12. c
13. a
14. b
15. b
16. a
17. c
18. False
19. c
20. a

Chapter 6
1. d
2. c
3. d
4. a
5. b
6. c
7. d
8. d
9. d
10. c
11. b
12. c

Chapter 7
1. d
2. a
3. d
4. c
5. b
6. c
7. b
8. b
9. b
10. a

Chapter 8
1. e
2. e
3. c
4. e
5. a
6. d
7. a
8. c
9. d
10. a

Chapter 9
1. b
2. a
3. c
4. d
5. b
6. e
7. e
8. e
9. d
10. d
11. c
12. a
13. b
14. d
15. d
16. b
17. d
18. d
19. d
20. b

Chapter 10
1. e
2. b
3. d
4. a
5. c
6. b
7. d
8. a
9. d
10. d

Chapter 11
1. d
2. d
3. a
4. a
5. c
6. b
7. b
8. c
9. d
10. c
11. a
12. c